THE HISTORY

OF

PUTNAM COUNTY, N. Y.;

WITH AN ENUMERATION OF ITS

TOWNS, VILLAGES, RIVERS, CREEKS, LAKES, PONDS, MOUNTAINS, HILLS, AND GEOLOGICAL FEATURES;

LOCAL TRADITIONS;

AND

SHORT BIOGRAPHICAL SKETCHES OF EARLY SETTLERS, ETC.

BY WILLIAM J. BLAKE, ESQ.,
COUNSELLOR-AT-LAW.

NEW YORK:
PUBLISHED BY BAKER & SCRIBNER,
146 Nassau Street, and 36 Park Row.

1849.

Notice

In many older books, foxing (or discoloration) occurs and, in some instances, print lightens with wear and age. Reprinted books, such as this, often duplicate these flaws, notwithstanding efforts to reduce or eliminate them. The pages of this reprint have been digitally enhanced and, where possible, the flaws eliminated in order to provide clarity of content and a pleasant reading experience.

Originally published
New York:
1849

Reprinted by:

Janaway Publishing, Inc.
2412 Nicklaus Dr.
Santa Maria, California 93455
(805) 925-1038
www.JanawayPublishing.com

2007

ISBN: 1-59641-125-2
ISBN 13: 978-1-59641-125-8

Made in the United States of America

PREFACE.

WE hardly know what excuse to offer the reader, for daring to introduce our little work into his literary presence, and lay his time under an embargo for its perusal. We have not written for fame or profit. Had we done either, or both, we would have selected a different subject than the history of a little County like that of Putnam. What we have written is the result of leisure hours, which might possibly have been squandered in the pursuit of a less worthy object, productive of no benefit to self or community. If we have garnered up one fact that was in danger of being lost, and which is beneficial and worthy to be preserved, in a historical point of view, we are satisfied and rewarded for our labor. That our little work is defective, no one is more thoroughly satisfied, and ready to admit it also, than our humble self. Our sources of information have been narrow and limited, and in many instances defective.

A generation has passed away; in this County, from whom valuable Revolutionary information might have been obtained. In addition to this, the records of the different towns in some instances have been defaced, by accident, we presume, and become obscure by lapse of time; rendering the record unintelligible and valueless to the delver in search of the treasures of the past.

We had intended to incorporate an outline view of Dutchess County in our little work, but finding that it would increase our pages beyond a given number, we were necessitated to withhold the article already prepared.

COLD SPRING, 1849.

THE

HISTORY OF PUTNAM COUNTY.

GENERAL VIEW OF THE COUNTY.

PUTNAM was erected wholly from Dutchess, June 12th, 1812; and was named in honor of Major-General Israel Putnam, who was stationed for some time, during the Revolutionary war, in the lower part of this county, and at Peekskill in the county of Westchester. It is situated on the east side of the Hudson river, between 41° 20′ and 41° 30′ north latitude, and 2° 56′ and 3° 26′ east longitude, from Washington. It is bounded northerly by the county of Dutchess, easterly by the State of Connecticut, southerly by the county of Westchester, and westerly by the Hudson river, which separates it from the counties of Rockland and Orange. Its area is about 216 square miles. Its population in 1840, was 12,825; and in 1845, 13,258.

It contains six towns, viz.: Philipstown, Putnam Valley, Southeast, Carmel, Patterson, and Kent. It was originally called the South Precinct of Dutchess county, and about 1740, the Fredericksburgh Precinct, embracing the whole of Putnam. As early as 1772, the present town of Philipstown, including Putnam Valley, was erected in a precinct, by the name of "Philipse Precinct;" and, in 1773, the town of Southeast was organized as a separate precinct, by the name of the "Southeast Precinct." This left in the Fredericksburgh Precinct, only the towns of Carmel, Kent, and Patterson. By the Act of March 7th, 1788, the terms precincts were dropped, and "Philipse Precinct," was called Philipstown; "Southeast Precinct," Southeast town; "Fredericksburgh Precinct," Frederick's town.

Philipstown is named in honor of the Philips family, a member of which patented the whole of this county; Frederick's town, in honor of the christian name of Capt. Frederick Philips, who inherited one third part of it, and Southeast, from its geographical position with respect to the other towns.

The geographical shape or figure of this county, is a geometrical rectangle, having its angles right angles, without having its sides equal. It stretches, like a garter, from the Hudson to the Connecticut line; being, in a straight line, about twenty miles in length, and twelve in breadth. Carmel and Patterson were organized in 1795, from Frederick's town. This left Frederick's town embracing only the now town of Kent, which name was given to it about this time, in honor of the Kent family. Patterson, in 1795, was organized by the name of "Franklin," in honor of the old revolutionary philosopher and patriot; but, in a few years thereafter, it was changed to Patterson, in honor of the family of that name, who were early settlers there. Putnam Valley was erected in 1839, by the name of "Quincy," after the town of that name in Massachusetts, wholly from Philipstown; but in 1840, the name was changed to the one it now bears.

The eastern part of the county is uneven and hilly, yet very productive, and under a high state of cultivation. The central and western portions are broken by high hills and mountain elevations. The Highlands stretch across its west end, casting their sombre shadows on the noble river, that laves its western boundary. They are estimated at 1,500 feet above the level of the Hudson. Through the central Highlands, run two valleys, called Peekskill and Canopus Hollows; and between them and the Hudson lies a beautiful vale, called Pleasant valley, extending from the Westchester to the Dutchess line. The mountain slopes and valleys are productive and well-cultivated. The Muscoot river, with the east and west branches of the Croton, are the only streams of any importance. There are several brooks and creeks, which furnish sufficient water power, for the milling purpose of the country. Iron is found in abundance in the mountains; and, though "bleak and barren, as appear these rock-ribs of earth, they are the repositories of exhaustless wealth, which requires but the hand of industry, to unlock and scatter to the world."

Extract from the record of the minutes of the first Court held in the county, after its organization.

"*October Term*, 1812.

"At a Court of General Sessions of the Peace, held at the Baptist meeting-house, in the town of Carmel, in and for the county of Putnam, Tuesday the 20th day of October, 1812.

"Present,

   STEPHEN BARNUM,
   ROBERT JOHNSON, } *Judges and Justices*
   BARNABAS CARVER,   *of the Peace.*
   HARRY GARRISON,

"Proclamation that Sheriff return precept, and other precepts returnable here this day.

"William H. Johnston, High Sheriff of said county, returns the ——, with the panel of Grand Jurors. The panel being called, the following grand jurors appear and answer, and were sworn, to wit, Joshua Barnum, junior, William Field, Jonathan Morehouse, Jesse Brush, Nathaniel Forgusan, Silas Whitney, Abijah Seely, Edward Vermilier, Joseph Cole, junior, Daniel Thorn, Benjamin H. Miller, John Austin, Tracy Ballard, Judah Relley, Jeremiah Hopkins, Ebenezer Boyd, Rowland Bailey, Joshua Hazen, Abel Peck, John Hyatt, junior, Abraham Smith, Jeremiah Conklin, and Amos Conklin.

"Proclamation, that all Justices of the Peace, Coroners, Sheriffs, and other officers, that have taken any inquisitions or recognizances, to hand them into Court, that the Justices of the people may proceed thereon.

"Proclamation, that all constables appear and answer. The list being called, the following constables appeared, and answered, to wit: Robert Post, Philips; James Randal, Frederic; Jersham Jacocks, Patterson; Lewis Baker, Southeast; Jesse Hill, Carmel.

"Proclamation for those who are bound in recognizances to appear and answer.

"Court adjourned till half after three, P. M. Court met pursuant to adjournment.

"Present,

STEPHEN BARNUM,
ROBERT JOHNSTON,
BARNABAS CARVER,
HARRY GARRISON,
} *Judges.*

"On application to this Court, of Abraham Smith, foreman of the Grand Jury, that they have no district Attorney. Ordered that G. W. Marvin serve as such, during this Session. Court adjourned, till to-morrow morning at 10 o'clock.

"*Wednesday,* 21st *October,* 1812.

"Court met pursuant to adjournment.
"Present, the same Judges.
"Court adjourned till 3 o'clock, P. M.
"Court met pursuant to adjournment.
"Present the same Judges.

"*The People vs. Moses Northrup.*"
"Indictment for an assault with intent to kill. Defendant pleaded not guilty. Ordered that this suit be suspended, till tomorrow morning, at 10 o'clock."

"*The People vs. Tartulus Russel.*"
"Indictment for an assault and misdemeanor."
"Court adjourned, till 10 o'clock, to-morrow morning."

GEOLOGY.*

As might be inferred, from the geological character of this county, its mineral productions are of much interest. Pursuing the order which has been heretofore adopted, it may be observed, that in iron ore, this county is peculiarly rich. It contains several beds or veins of the magnetic kind, which yield ore of the best quality, and in the greatest abundance.

Of carbonate of lime, in the form of calcareous spar, and marble, there are several localities. The latter term, however, is usually applied here, to a dolomite, similar to that of Dutchess county, and which is found abundantly, in the vicinity of Patterson. The only objection to this material for construction, is its friable character.† Being a compound of the carbonates of lime, and magnesia, it has been thought that the product of its calcination, is not so valuable, as a fertilizing agent, as that which contains lime alone. On this subject, however, as I have

* Dr. L. C. Beck's Geological Report, 1839.
† The Putnam county dolomite, often contains a large admixture of tremolite.

already remarked, mistaken views have been entertained, as some soils which appear to have been formed in part, of the magnesian limestone, are by no means wanting in fertility. It may be added, that a white crystalline limestone occurs in this county, which is entirely free from magnesia.

A very valuable mineral product of Putnam county, is serpentine, which exists in vast quantity, can be obtained in blocks of uniform density, and is susceptible of a fine polish. But for the fact, that the quarries of this beautiful material are situated at too great a distance from water transport, they would long ago have been extensively wrought. Of the precious or noble serpentine, also, there are several localities, where the mineralogist may obtain specimens, equal in beauty to any that are found in the United States. Among the objects worthy of notice in this connection, are two localities of arsenical iron pyrites, one of which has, at some former period, been extensively wrought. This ore, which might be used for the extraction of arsenic, seems to be abundant, but the present condition of the mine renders it difficult to arrive at a certain conclusion on this subject. Its reported mixture with silver, is probably one of those stories circulated for selfish purposes, and the only ground for it, in the present instance, is, the fact that in other countries, the arsenical ores are often associated with those of a more valuable metal.

I should not omit to add, that a bed of limonite or hydrate of iron, occurs in Peekskill Hollow, near the line between Philipstown, and Carmel. Mr. Mather, however, remarks that it seems to be too silicious to work well alone, in the furnace. There are also several localities of peat, and probably marl will hereafter be found accompanying this substance, as it does in several other counties. Putnam county has, for many years, been visited by mineralogists chiefly on account of the interesting localities, which occur at Cold Spring and in Philipstown. The latter on the farm of Mr. Huestis, about five miles south of the village of Fishkill, is particularly worthy of attention, not only for the number of interesting minerals which it affords, but as showing the manner in which allied species run into each other. The facts here presented, prove how important it is in many cases, carefully to examine the localities of minerals, before we decide with certainty upon their specific characters.

In a bed of white limestone, running parallel with the granite, and which is of small width, and is situated about a quarter of a mile from the New York road, we have the following minerals, viz.: precious serpentine, of which there are several varieties. The first has a conchoidal fracture, and presents various shades of green and yellow, and is variously disseminated through the limestone, and other minerals. 2. A slaty variety, having a dark green color. It sometimes breaks into rhomboidal prisms, and is very hard and compact. 3. A slaty variety of a greenish white colour, is harder than the preceding, and is fusible upon the thin edges by the blow-pipe. This last seems to approach jade or Saussurite in its characters, but its peculiarities are probably owing to the admixture of some other minerals, as all these varieties seem to pass into each other by almost imperceptible gradation.

Of the magnesian carbonate of lime, several forms occur at the locality in question. There is a thin stratum which is snow-white, very close grained, compact and has a semiopaline appearance. It is sometimes described under the name of Gurhofite, from its having been first found near Gurhof, in Austria. According to my analysis, its composition is as follows, viz.:

Silicia,	6.50
Carbonate of lime,	66.75
Carbonate of magnesia,	26.50

Other specimens of this mineral, have fibres of asbestus running through them, and they sometimes have a bluish tint, a slaty structure, and contain crystals of bronze yellow iron pyrites.

Asbestus, especially in the amianthoid form, is, indeed, largely mixed with the minerals already noticed. The beautiful silky fibres, which run through the serpentine and magnesian limestone, in various parts of this ridge, may belong to the picrolite of the more recent authors. But, while I have no doubt, that several distinct minerals have heretofore been confounded, under the name of asbestus, I have not satisfied myself of the identity of our mineral with that just mentioned. Some of the specimens found here, have a fibrous structure, and a silky lustre, but on treating them with acids they effervesce, and after dissolving out all the magnesian carbonate of lime, bundles of amianthoid fibres remain. I have reason to believe, that all the different fibrous minerals found here, are mixtures of asbestus with serpentine or magnesian carbonate of lime.

The following, is an enumeration of the remaining minerals found in this county:

Pyroxene.—The variety coccolite, of a white color. It is made up of grains of various sizes, and resembling dolomite—Philipstown. *Diopside*, or white augite, associated with the preceding. It has a grey color, and foliated structure.

Tremolite.—Often associated with yellowish serpentine.

Chrome iron ore.—In octahedral crystals. One specimen has a crystal with the base one-fourth of an inch in diameter. It resembles spinelle, but its power is attracted by the magnet, and its blow-pipe characters agree with those of the chrome ore—Philipstown.

Scapolite, in small crystals or grains. Phosphate of lime, in very minute crystals. Both these are found at the above locality.

Peat.—Several localities.

Graphite.—Often found in small quantities.

Arragonite.—In concretions on dolomite, near Patterson.

Hyalite.—Associated with iron ore, at the Philips ore bed.

Basanite.—Loose masses on the banks of the Hudson.

Kerolite.—Philipstown.

Brucite.—Near the Townsend ore-bed.

Hornblende.—Several varieties.

Actynolite, found at Brown's quarry. *Tremolite.*—Very abundant in the dolomite near Patterson. *Asbestus*, and *amianthus*, in long and tough fibres. Abundant near Patterson.

Schiller spar, of Dr. Thompson. Brown's quarry.

Feldspar.—Several varieties.

Albite, in large crystals, near Patterson.

Laumonite, Stil ite, and Chabusie.—Formerly obtained at Cold Spring. The locality is believed to be exhausted.

Epidote, in beautiful crystals, near Carmel.

Mica.—Several localities.

Zircom.—Formerly obtained at Cold Spring.

Iron Pyrites.—Associated with magnetic iron ore. By long exposure to the weather, the iron pyrites is decomposed, and the resulting salt washed out.

Pyritous Copper, and Green Carbonate of Copper.—Philips' ore-bed.

Sphene.—At the Philips' ore-bed, and formerly at Cold Spring.

Orpiment, or Yellow Sulphuret of Arsenic.—Formed on the timbers of the old arsenic mine, by the decomposition of the arsenical iron pyrites.

Copperas, or Sulphate of Iron.—Formed by the decomposition of iron pyrites, on the farm of J. Wood, six miles S. S. E. of Carmel.

Schiller spar, or Metalloidal Diallage.

As there is some confusion in regard to the above names, it is proper to state that the mineral about to be noticed, is identical with the Schiller spar of Dr. Thompson.*

Color: dark green, almost blackish green. Fracture: uneven, splintery. Sectile. Specific gravity: 2.746. It is in broad, foliated masses, which cleave in two directions, and apparently have the primary form of a rhombohedron. The lamina slightly curved. One of the cleavages is easily obtained, and has a metallic, pearly lustre, and a pinchbeck brown color.

* Outlines of Mineralogy and Chemical Analysis, I. 173.

Hardness about the same as that of serpentine. Powder, yellowish gray. Where the mineral has been exposed to the air, it is of a tombac brown color.

Thin fragments treated by the blow-pipe are merely rounded on the edges, but become of the same brown color as when they have been long exposed to the air, and are attracted by the magnet. With borax it is fusible, though with difficulty, and the glass, when cold, has a greenish color.

This mineral is found associated with dark-colored, common serpentine, at Brown's quarry, near Carmel, Putnam county. Its composition is no doubt influenced by its contact with the latter substance.

*Sulphate of Lime, Alumina, &c.**

In the Highlands, sulphate of lime is frequently seen incrusting hornblende and augitic rocks. One kind of hornblende rock is common, that is more or less distinguished by this mineral. It is formed by the decomposition of pyrites; and the acid combining with the lime and crystallizing, causes the rock to crumble. These masses, when imperfectly crumbled, generally have a yellowish and reddish-brown color, from the oxide and red sulphate of iron, while the interior of the mass is filled with thin plates and crystals of the sulphate of lime. It is very common about West Point. Acicular sulphate of lime, very beautiful, was found incrusting augite rocks which overlaid white limestone, at an old mine-hole on Anthony's Nose mountain, about three miles east of Fort Montgomery, near the top of the mountain. Magnetic pyrites occur both in the augite and limestone.

Acicular sulphate of lime was seen incrusting mag-

* W. W. Mather's Geological Report, 1843.

netic oxide of iron, at a mine where pyrites abound in that ore in the Philips vein, eight and a half miles from Cold Spring, on the road to Putnam court-house, in this county. The specimens were very delicate and beautiful.

In this county, there is a locality of sulphate of alumina and iron, and of sulphate of iron, in Philipstown, on Anthony's Nose mountain, about three miles from West Point, at an old iron mine, where the ore contains pyrites. The earth from this place was used many years since, by some of the inhabitants, for dyeing. Another locality is near Luddington's Corners, half a mile east, in Kent; another in the same township, four or five miles south, near Dean Pond.

Sulphur and Carburetted Hydrogen.

In Philipstown, along the shore of the Hudson, south of, and near, the point of Gouverneur's Cove, about east of Gee's point, where pyrites have decomposed, a gray or bluish-gray powder, composed almost entirely of sulphur, is found. Another locality is about one mile east of the above, near the locality of laumonite and stilbite: it is in cavities in quartz, where pyrites have decomposed.

Sulphate of Iron.

In this county, many localities of sulphate of iron were observed.

1. An old mine-hole in hornblende rock, a quarter of a mile east of Luddington's Corners, in Kent, six miles north of Carmel village. Copperas effloresces, and causes this rock to crumble to sand. The excavation, which is small, is on the west side of the mill-pond.

2. In a vein of quartz, about thirty rods east of Deau Pond, also in Kent. Several metalliferous minerals are found here and in this vicinity.

3. About one-fourth to a half of a mile south-west of Pine Pond, in Kent, at the mine of arsenical iron. The ore, which is abundant, and situated in hornblendic gneiss rock, decomposes on the surface, forming sulphate and arseniate of iron. The mine is called the *silver mine*, and silver is said to have been obtained from the ore. If it contains this metal, Prof. Beck will make it known in his report containing the analysis of the minerals.

4. Another locality was observed two miles west of the last, half a mile to one mile from Boyd's Corners, on the road to Cold Spring, near the turn of the road, exposed in digging the road. This was arsenical iron, decomposed on the surface into sulphate and arseniate of iron.

5. At one of the excavations for magnetic iron ore, eight and a half miles from Cold Spring in Philipstown, on the road to Carmel village, and half a mile, or perhaps more, north-east from the principal of the Philips mines, on the same vein, five or six hundred tons of ore have been blasted from the vein. At this locality, the magnetic oxide of iron is so much intermixed with pyrites, that it cannot be used to make iron. In some places in the vein, the pyrites seems to have been a paste in which the grains of magnetic oxide of iron have been disseminated, but it does not generally form more than one-fourth to one-sixth of the mass of that part of the vein. By exposure to the weather, copperas is formed, which effloresces in dry weather, is washed away by the rains, and is succes-

sively formed and renewed, until the pyrites is decomposed and the magnetic oxide left nearly pure.

6. It is believed that sulphate of iron might be manufactured at this place for the market. A slight roasting would facilitate the decomposition of the sulphuret.

7. In Patterson, several localities were observed where pyrites decomposed and formed copperas. One is near Mr. Robinson's farm, four miles north-east of Carmel.

8. Another on the same road to Patterson, and within two or three miles of the village, on the high ground. The gneiss, about one hundred or two hundred yards west from its junction with limestone, contains much pyrites. The metalliferous bed seemed to be five or six feet in thickness in the vertically stratified rock, and sulphate of iron effloresced on the surface

9. Another locality, about a mile west of Patterson, is in a ridge of gneiss, between strata of limestone. This pyrites gneiss stratum extends a distance, it is supposed, of several miles, and may at some time be used for the manufacture of sulphate of iron.

10. In South-east, a locality of sulphate of iron was examined on Mr. Jedediah Wood's farm, six miles south-east of Carmel, on the hill west of the Croton river. Pyrites abound in the rock on the eastern brow of the hill, and copperas effloresces on the face of the rocks. Many of the loose masses at and near the foot of the hill are porous, as if once filled with pyrites, which have decomposed and washed away. Excavations have been made in two places where the pyrites abound. Some of the rock crumbles by the disintegrating action of the crystallizing salt. The

pyritous stratum was traced along the brow of the hill about two hundred yards. It has long been supposed there was a lead mine in this hill; and perhaps it may not be inappropriate here to mention circumstances that serve to give countenance to that idea among some people, who still yield implicit faith to the miraculous virtues of the "mineral rod" and magic glass, when used by the favored few who pretend to be gifted with such peculiar powers.

It is stated by various gentlemen of the highest veracity, that a man of high respectability came from Connecticut to the owner of the farm, and informed him that there was a valuable lead mine on his land, which was worked many years before; that it was covered over with planks; that a walnut stick was lying with one end on the planks, and the other mostly decayed near the surface; that the earth had washed over it about four feet deep (with other circumstances of detail), and that he could go directly to it. He would make no communication to indicate the locality, until the owner had executed a bond to secure a certain portion of the profits to the informer, who had never been to the locality, but who stated that he was enabled to see it, and go directly to it, by looking at a polished stone as a mirror placed in the crown of his hat, with his face applied to the opening of his hat, to exclude the view of other objects. He seemed a perfectly unassuming, quiet man, with a perfect faith in his ability to perform what he stated. Many of the citizens of the vicinity accompanied him in his wanderings, and he finally stopped near the foot of a hill on which we have described the pyrites, and where he directed excavations to find the mine. The walnut stick was found as indicated, except that there were no planks; and no opening, no trace of a mine, or of any ore, could be discovered. He went away much mortified with his failure.

A few years afterwards, a girl who was reputed to be able to see in a magic glass, or polished stone held in a dark place, was employed to discover the supposed mine, and it was said she had never been in the vicinity before. She looked, and walked to within a few yards of the same spot, drove a stake in the ground, and said the mine was *there*, at a depth of thirty-five feet; but none has been found.

A man who had moved into another part of the country when a boy, returned when old, about thirty-five years ago, and stated, that when he was a boy he had been into a mine in that hill where the lead ore had been dug, and that he hád seen the ore. He had endeavored to find the mine, without letting the people know the object of his search. Hearing these stories, and many believing that there is a lead mine in that hill, led me to make an examination of the locality with as much care as was practicable without excavations. The loose stones on the surface are more or less porous gneiss, with a reddish color. Both the porosity and color are due to the composition of pyrites probably, for I discovered no traces of any metal but iron. These appearances would very naturally induce the idea of a mine, even without the aid of a *magic glass;* but whether any ore of any value occurs *there*, is a subject for investigation. There are no indications that seem to justify the expenditure of capital in search of lead or other metals. The rock near the brow of the hill abounds in pyrites, is nearly vertically stratified, ranges north-north-east and south-south-west, and is intersected by quartz veins (that show no metallic contents) in a south-east and north-west direction. The great vein of magnetic oxide of iron called the *Simewog vein* (in consequence of the extensive mines of this ore that have been worked in Simewog hill), is about one-eighth of a mile west of this locality of pyrites, as is supposed from the great variation of the compass near Mr. Wood's house. The compass varies in a distance of two rods in an east and west line from 30° to 40°, and the centre is in the north-north-east and south-south-west line of direction of the Simewog vein on Simewog hill. On the north-north-west side of where the vein seems to pass, the compass varies to the east of north 15° to 20°; and on the south-south-east side, it varies as much to the west of north. The stones on the surface where the vein is supposed to pass, contains magnetic oxide of iron disseminated; and some lumps of ore, and masses of magnesian garnet and epidote, were also observed on the same line at this place, and a mine of the magnetic ore has been opened about a mile south-south-west. The white limestone is not far distant on the east from the reputed lead mine.

All the geological circumstances that were observed, and that seem to have a bearing upon this reputed lead mine, have been related. The region is a highly metalliferous one, and it is probable some excavations may have been made here, as in hundreds of other places in the Highlands, by the company of miners that was sent into this country between 1730 and 1750, under the direction of the Baron *Horsenclever;* and that traditions connected with these excavations, aided by some imagination and credulity, have been the basis of many of the reputed silver and lead mines in the Highlands and other parts of New York.

Many localities might be mentioned in this county, where pyrites decomposed with the formation of the sulphate of iron. The principal that have not been mentioned, are,

1. An old iron mine on Anthony's Nose mountain, about one and a half miles east of Fort Montgomery.

2. An old "silver mine" (but which contains no silver), on the top of the same mountain, three miles east of Fort Montgomery.

3. On the top of a hill, about one and a half miles east of West Point.

4. Shore of the Hudson, nearly opposite Buttermilk Falls.

5. Shore of the Hudson, several places, nearly opposite West Point.

6. Shore of the Hudson, several places, between Arden's landing and the landing above.

7. Near the post road, two or three miles east of Arden's Landing.

The localities in this county are all in the primary rocks, and nearly all are at or near fractures or uplifts, or localities of more than usual disturbance of the strata.

Altered Taconic Rocks through the Highlands.

In crossing the Highlands, the Taconic rocks are more or less altered. Some of the limestones in the towns of Beekman and Fishkill, in Dutchess county, where they approach the Highlands, are white and grey crystalline limestones.

The *granular quartz rock* forms a continuous stratum through a portion of this and Westchester county. It is probably a continuation of the stratum described as forming a part of Peaked and Elbow mountains in Amenia and Dover in Dutchess county, and is probably a continuation of that described by Prof. Hitchcock in the western part of Massachusetts.* The granular quartz rock crops out on the bank of Peekskill bay of the Hudson river, about half a mile north-west of Peekskill landing, near Hall's point. The strata are nearly vertical, leaning a little to the west-northwest. It ranges up the "Peekskill hollow." It is seen in connection with the iron ore at Bradley's ore bed in Peekskill hollow, about ten miles from Peekskill; and again it occurs near Boyd's corners, in Carmel, in this county. It is quarried to a small extent near Boyd's corners, for door-steps, hearth-stones, and other purposes. It splits out in regular slabs from three to nine inches thick, and three to seven or eight feet square, with an uniform plane surface, and is admirably adapted for a flagging stone for streets, cellars, &c.

The locality near Hall's point belongs to ——, of Peekskill. It is at the mouth of Peekskill creek, a little north of Hall's point; and it is believed that a valuable quarry of flagging stones, of the granular quartz rock, may be opened at this place. The strata are nearly vertical, and the stones may be split off with great ease, if the quarry be opened in a proper

* Vide Second Annual Report on the Geological Survey of N. Y., p. 172; Hitchcock's Geological Reports of Massachusetts, 1833, pp. 22, 321; Hitchcock's Final Report, 1842, pp. 587, 593; Dewey's American Journal of Science, Vol. 8, &c.

manner. The flagging and curb stones used in New York, and many other towns, are now brought from the Bolton and Haddam quarries in Connecticut, and from the Greywacke quarries in Greene county. The Bolton and Greene county stone are carted from eight to sixteen miles over bad roads, and then shipped to a market. If stone, as beautiful and as durable, can be dug on the shore of the Hudson, where no cartage is required, and where the expense of quarrying is no greater than at the quarries mentioned, and where the business is now very lucrative, it follows that such quarries on the shore of the Hudson would be very valuable.

The *talcose slate*, distinctly characterized, is limited in extent. It occurs in this and Westchester county, forming a range of hills several miles in length. It forms Blue-rock Point, on the post road, between the crossing of Peekskill creek and Annville. The slaty laminæ are parallel in direction to the limestone and granular quartz rock on the east, which dip at an angle of from seventy-five to eighty-five degrees to the east-south-east. This rock forms the principal mass of the hills to the north-north-east of Blue-rock Point for several miles. Gallows Hill (a place celebrated during the Revolution, in consequence of the public executions), is a part of this range of rock. The rock is generally covered by soil, except where it has been denuded by water, or excavations for roads, etc. The soil is of good quality, and produces fine crops. Farther north-east, this rock is rarely seen; but it passes up Peekskill hollow, and up a valley two or three miles west of Boyd's Corners in this county. It is very refractory in the fire, and is used for the in-walls of furnaces. The rock is very fissile, and splits in thin laminæ of some magnitude.

The limestones of this range of Taconic rocks are scarcely altered in some places, as at the quarries near Blue-rock Point, near the mouth of Peekskill Creek, and about a quarter of a mile east of Annville, in Westchester county. In others, it is a perfect metamorphic white limestone, as in the valley west of Gallows Hill, two or three miles north of Annville.

Limestone makes its appearance as knobs or hills, fifty to one hundred feet high, about two or three miles north of Annville, in the valley west of Gallows Hill. Also the same limestone, in the same valley, associated with talcose rocks, two or three miles north of the last locality, near Bunnell's forge. The strata are nearly vertical.

Metamorphic Limestones.

In tracing these limestones in Dutchess county, we left them in Pawlings. The same granular dolomitic limestone extends south into Patterson, where it is well exposed to view, from the north line of Patterson to three or four miles south of the village. It is associated with mica slate, and a fissile, micaceous gneiss rock. The limestone in the valley of Patterson contains tremolite in some places. About two and a half miles south of Patterson, the limestone is quarried for lime, and forms a superior article. Sixty cords of wood are consumed in burning a kiln of two thousand bushels. The price of this lime is fifty cents per bushel.

Much of this stone seems well adapted for a building stone. The rock is granular, strongly coherent, and in color varies from bluish to white. The rock is not fitted for a marble, as the tremolite would make it difficult to saw.

Another quarry has been opened within a mile of the village of Patterson.

About a mile west of Patterson, a ridge of gneiss, highly impregnated with pyrites, lies between strata of limestone. The strata through this region dip to the eastward nearly vertically.

Limestone of the same general characters occurs in Southeast, in the valley of the Croton river. It may be seen along the road about a mile above Owensville, and one mile and a half west of Peach Pond.*

Serpentine Rock.

Another locality of this rock is in Philipstown, about ten or eleven miles north-north-east of Peekskill, and about half or three-quarters of a mile east of Horton's Pond. The rock is of a blackish-green, fine grained, and sometimes coarsely crystalline. It is yellow on the weathered surfaces, and is associated with steatite. Ten to eleven acres seem to be underlaid by this rock, which might be quarried for an ornamental marble. It is about eight miles from water transport. Another

---

* I may be permitted here to mention a circumstance respecting *ground ice*. The ice in this pond attaches itself to the loose boulders in shallow water, and floats with them attached, and ploughs up the gravel before them as the ice is driven towards the shore in the spring. In this way they are brought annually nearer and nearer the shore, until they are pushed beyond low-water mark, where they remain. Many of these boulders weigh fifteen or twenty tons. They all seem to have come from one general direction, viz., north-west. Mr. Stephen Ryder pointed out the boulders and furrows, and gave the explanation to Prof. Cassels.

mass of the serpentine rock was seen about a mile south of the one last described. The serpentine forms knobs and hillocks at small intervals for half a mile in length. Another locality, one that has already attracted much notice, is Brown's quarry, near Pine Pond, in this county, four or five miles from Carmel village, and one and a quarter miles north-north-west of the county poor-house. It is dark-colored, dark green to black, and from compact to a coarse crystalline, like coarse-grained hornblende rock. It is granularly foliated, like common white marble, polishes well, and is perfectly black when polished. It may be obtained in large blocks for sawing into slabs. Large blocks lie on the surface in Brown's lot, and the rock is seen in place all around the hill. In the mine lot adjacent, good blocks may probably be obtained by quarrying. Twenty-five to thirty acres of ground are underlaid by this rock on the hill-side, west of the brook, which is the outlet of Pine Pond. It is easily accessible, and about one hundred feet above the water level of the adjacent valley. Blocks of many tons' weight can be easily procured; in fact, many of this size are now lying on the surface, and require no blasting or splitting before they are put in the saw-mill. Magnetic oxide of iron, or chromate of iron, is disseminated through the serpentine in some parts of the serpentine bed; and this variety of the rock will not be suitable to work, as it can neither be sawed nor polished easily. The quarry seems to be sufficient to supply the market, not only of our own country but the world, with this kind of ornamental marble, for a long time. It is really a beautiful material when polished, and it is hoped that it will be extensively used.

I have seen no other locality where such a material can be obtained in so large blocks, sound and free from seams and cracks. A marble of this kind was used in ancient times, in some of the old Spanish palaces, but it is exceedingly rare in Europe.

Metamorphic Limestones of the Highlands.

These are similar to those in Orange county, only spinelle has not been recognized in them, and brucite is not common. Serpentine, augite, and asbestus are more common, and garnet is more common in the associated rocks.

Local Details.—1. Limestone was observed about one and a half miles south of Carmel village, on the farm of a Mr. Townsend, at two old mine holes, where some have supposed that silver, and others that marble was the object of exploration. It is scarcely necessary to add, that no traces of silver ore could be distinguished. Both these excavations are in a bed of limestone, about thirty rods apart. The bed is narrow, perhaps twenty feet wide, and is bounded by gneiss on each side; the strata are highly inclined to the east-south-east. Brucite and some coccolite were observed in the limestone of the northwardly excavation. At the other locality the limestone is very white, coarse-grained, and contains imperfect crystals of phosphate of lime, or green augite.

2. A bed of limestone containing brucite, serpentine, and asbestus, is associated with the bed of magnetic oxide of iron on Mr. Tilly Foster's farm, two and a half miles south-east of Carmel village.

These were the only localities of this kind of limestone seen in the *eastern* part of this county. In the *western* are two ranges of the same kind of rock. The following localities will illustrate them, viz.:

1. A bed of limestone near Anthony's Nose point, six miles south of West Point, which is seen again farther north-north-east at the old silver mine; also at two places on the shore between that and cotton rock; near Mr. Arden's; two hundred or three hundred yards south-west of Philips's mill, east of West Point; also at Cotton Rock; half-a-mile east of the Highland school; north of Davenport's tavern on the post road; and at Huestis's quarry. This is the westernmost range, and extends from near Anthony's Nose point north-north-east, to near Davenport's tavern; and thence through the valley to Huestis's quarry, about three miles south of Fishkill, a distance of about ten or eleven miles. It has also been seen farther to the north-north-east on the mountain.

2. The second range has not been observed in as many places. It has been seen at the White mine on the Anthony's Nose mountain, three miles east of Fort Montgomery; three miles east-south-east of West Point, near the post road; on the road from Cold Spring to Carmel village, near Haight's tavern, five miles north-east of Cold Spring; and about half a mile north of Warren's tavern, in Philipstown. This range is nearly parallel to the other, and nearly similar.

All these localities are supposed to be in the range of limestone which are exposed in these various points, and probably in many others.

1. Huestis's quarry is in the western limestone bed above-mentioned. Some parts of the hill are granular limestone, and a part is nearly compact magnesian

limestone or miemite. Serpentine is frequently intermixed, forming a verd-antique marble, which may perhaps at some future time be applied to use. Several fine minerals occur at this locality, which were discovered by Dr. Barrett in 1822. The precious serpentine of this locality is perhaps not surpassed in beauty by that of Newburyport or Easton, or even any locality known. It occurs crystallized distinctly with various modified forms. White coccolite, white augite, diopside, sahlite, phosphate of lime, amianthus, asbestus, pearl spar, pyrites, chromate of iron, magnetic oxide of iron, and various other minerals occur at this locality. The mass of limestone at this place forms a bed twenty to fifty feet thick, resting against granite or sienite, and gneiss containing red felspar and some epidote, on the west, while a stream flows at the base of the hill.

2. The bed of limestone north-west of Davenport's corners, which is on the post road five miles northeast from Cold Spring, is similar to that at Huestis's, but not as beautiful, and does not exhibit the beautiful minerals of that locality.

3. This range of limestone crosses the road about one hundred or one hundred and fifty yards west of the locality of laumonite and stilbite in Philipstown, about three hundred yards west of Philips mills, and one mile and a quarter east of West Point. It is white, and highly charged with grains of magnetic oxide of iron. Granular hornblende, like coccolite, also occurs with the limestone. Hornblende gneiss, and micaceous gneiss containing pyrites and scales of plumbago, also occur associated.

4. This bed has had excavations made in it, in two

or three places, between the locality above described, and where it reaches the shore of the Hudson about one mile and a quarter south-west of Mr. Arden's; and every place where it was examined, showed imbedded grains of magnetic oxide of iron, and in some places imperfectly characterized brucite and serpentine.

5. The "Cotton rock," as it is called, is at the junction of this bed of limestone with diallage and hornblende rocks. The limestone is extremely impure from contained minerals, so that it would scarcely be recognized as limestone from its external aspect, when weathered by one not familiar with all the protean forms of this rock; the calcareous matter being removed to some depth, and the imbedded mineral substance remaining to form a crust of some lines or even inches in thickness exterior to the sound, unaltered rock. Granite occurs in the cliff a few yards to the east, which contains imperfect crystals of black tourmaline.*

6. An outheaved mass, as it is supposed, of this range of limestone, occurs about a mile north of the last locality, near the Beverly dock (the point from which the notorious and treacherous Arnold made his

---

* The Cotton rock is a noted place in the Highlands. It is on the shore of the Hudson, in Philipstown, between high and low water mark, about three and a half miles below West Point. The rocks here are impure verd antique (composed of limestone and serpentine, with brucite and other minerals imbedded), serpentine with diallage, and veins of asbestus, and delicate silky amianthus, diallage rock, augite rock, and granite. The amianthus which comes from the veins in the serpentine, has the appearance of cotton or raw silk, and has given rise to the name Cotton rock.

escape). This bed of limestone, which is small, lies transverse to the general direction, viz., north-west and south-east. The limestone is colored with serpentine, and contains plumbago; but some of it is of a pure snow white, strongly translucent, and varies in texture from coarse crystalline to compact, like the finest alabaster. The white is much intermixed with augite and quartz, so that specimens for the cabinet can· be obtained showing the characters indicated, but not large masses. The associated rocks are micaceous pyritous gneiss, and grey augite containing hexagonal plates of plumbago. A vein of pyrophyllite, six inches to a foot wide, traverses the grey augite rock.

7. Some distance below the Cotton rock, perhaps one quarter to half a mile, the limestone bed that we have been tracing forms the shore for some rods. The same minerals are found here as at many other parts of the bed, viz., magnetic oxide of iron, serpentine, and augite and hornblende in the vicinity, associated with granite and gneissoid rocks. A little farther south the limestone has an old mine hole in it, a little south-east of a small bay of the river shore, where the rock contains much magnetic oxide of iron.

8. From this place, the limestone ranges south-southwest towards Anthony's Nose point. This part of the range is perhaps of more interest to the geologist than any other portion, and particularly the tract of land between the " *Old silver mine,*" as it is called, and the Hudson river, about a quarter of a mile in width. The limestone and all the associated rocks here, have been subjected to metamorphic agency in a high degree.

The limestone lies at the base of the hill, adjacent to a granite rock. It is white, highly crystalline, contains much mica, plumbago, and various mineral substances. The plumbago is generally diffused in minute particles, but in some places is so abundant as to give a bluish tinge in streaks through the rock. The resemblance of this mineral, thus diffused, to the sulphuret of silver in minute particles, its softness and lustre, led to the idea that silver ore was abundant at this place.

This mine was opened in ancient times, for what purpose is not known. Stalactites are found in the adit level, which is two hundred yards long, and old tools with the handles rotted out have been found in it. It was re-opened some years ago, with the hope of getting silver; the man who worked it having been told that the scales of plumbago in the rock were sulphuret of silver. He is said to have spent all his property, and to have died in a mad-house. Spene and zircon are occasionally seen in the augitic and calcareous rocks at this place, but they are not common.

The most interesting mineral substance in a geological point of view, found at this old mine, is quartz. It is strongly translucent, almost transparent, like hyalite, in irregularly round masses from the size of filberts to that of an egg. They seem to have been melted to assume their present form. The first public notice of such facts, so far as I know, was by Prof. Emmons, in the second Annual Geological Report of New York, 1839, p. 202. I had observed the rounded, apparently fused quartz at the opening next the marsh, near the south end of the adit level of the

"Old silver mine," in 1825; but did not consider it of any special importance as connected with the age of these rocks, until the subject was brought before the public by Prof. Emmons. The spinelles of some localities in Orange county have their angles rounded, and contain spherical cavities, apparently produced by the same cause; and the crystals of phosphate of lime, sent me by Dr. Crawe and Prof. Gray, from St. Lawrence county, have their angles rounded, and contain cavities which seem to be referable to the same agency that has caused the crystalization of the limestone, and the formation and crystalization of the plumbago, and various minerals in these rocks. The rocks at this locality of the "Old silver mine," are well worth studying. The rock next the limestone on the west of the adit, is composed of augite and manganesian garnet; sometimes one, sometimes the other predominating. Both are crystalline, and sometimes exhibit perfect crystals. The garnet and augite frequently assume the granular form of colophonite and coccolite; red for the first, and green, brown, and purple for the latter.

The rock is very heavy, and very similar to some of the beds at Rogers's rock on Lake George. The rock farther west, is a gneissoid silico-felspathic rock, containing in many places magnetic oxide of iron in grains, and in small strings and veins parallel to the strike of the rock; also schorl in masses of quartz, and sometimes crystals of allanite like those near Fort Montgomery.

Still farther west is a range of limestone, very similar to that at Cotton rock, but purer. It contains an abundance of imperfectly characterized brucite.

Gneissoid rocks intervene between this bed and another of limestone seen still farther west, which has the general characters of verd-antique, being composed principally of limestone and serpentine; but it contains other minerals that are hard, and would prevent its being sawed easily. This rock is at the south point (called Marble point), and a few rods south of the north point of this neck of land. Granite is frequently seen interlaminated among the strata described, which are about vertical. The gneiss along the shore between the two last masses of limestone, and west of the last mass described, along the shore, contains pyrites, and by its decomposition, gives a reddish tinge to the rocks. The same character, and arising from the same cause, is seen in the slaty and micaceous gneiss along the shore, most of the distance from this place, north-north-east to Gouverneur's landing opposite West Point.

The *second range* of limestone, a mile or two east of the one we have been describing, ranges about parallel, and nearly as great a distance.

1. The " White mine," as it is called, is the most southern point at which the limestone of this range, as it is supposed, was seen. It is on Anthony's Nose mountain, about three miles east of Fort Montgomery. The limestone is white, highly magnesian, and contains some carbonate of iron, and perhaps might without any impropriety be called brown spar. It is crystalline and sometimes compact, and contains granular magnetic oxide of iron. It is associated with augite and granitic rocks.

2. Another locality of this rock is about two miles to the north-north-east, near the old post road, at .

reputed lead and tin mine. The rock is limestone with some serpentine imbedded, and contains grains of the magnetic sulphuret of iron. Near this place, where the lead is said to have been formed from the ore in the soil by burning brush-heaps, the soil is red, as is so commonly the case where the calciferous sandstone has been upturned and partially altered. The same limestone is seen still further to the north-north-east, on the hill, but was not particularly examined. The same silico-felspathic gneissoid rocks, and pyritous micaceous gneiss, as described near the "Old silver mine," occur in this vicinity, and the rocks are in some places very much confused.

3. The limestone near Haight's tavern, is similar to that described above.

Steatite.—A locality was seen near Peckville, a little north of the line of this county, and within Dutchess county. It is there' intermixed with serpentine, and although abundant, and quarried in large blocks, it was found difficult to saw it well, in consequence of the different degrees of hardness of the steatite and serpentine. It is beautifully spotted and clouded; and as steatite indurates by heat, it is possible that it may at some future time be wrought as an ornamental stone. Some of the masses of steatite are very pure, soft, and easily wrought. In some parts of the bed the rock is granular, or scaly talc, either pure or traversed in every direction by crystals of actynolite.

Another locality was seen in Philipstown, in this county, on Mr. McCabe's farm.* It is near the serpentine rock before described as eight or nine miles north-north-east of Peekskill, and half to three-quar-

* James McCabe, Esq.

ters of a mile east of Horton's Pond. The rock here graduates through every variety of aspect, from talc, through steatite, to serpentine. I did not see proper soap-stone or steatite rock adapted for useful purposes, *in place ;* but was assured that large blocks had been dug there, and that there was an abundance of it. I saw slaty, steatitic rock *in place*, and small masses of beautiful steatite scattered over the ground. Good quarries of this rock are well known to be very valuable. The blocks are worth twenty dollars per ton in market. This bed graduates on the east into serpentine rock.

Limonite, or Hematite Ore-beds.

Bradley's Ore-bed.—A bed of limonite, containing some oxide of manganese, occurs very near the line between Philipstown and Carmel, in Peekskill hollow, about ten miles north-east of Peekskill. Its thickness is not known. Fifty to one hundred tons of the ore may be seen in heaps on the ground. The ore was dug many years ago, but it seems to be too silicious to work well alone in the furnace. By proper mixture with other ores, it might be wrought with advantage. Isaac Lockwood owns a part of the land underlaid by this ore. The right of digging the mine is vested in Nathaniel Bradley, of Connecticut, who purchased a large amount of mineral property in the Highlands some years ago. The ore is associated with granular quartz on the east, and probably with limestone on the west, but this latter rock was not seen near the ore beds.

These rocks are associated in the above order at the mouth of Peekskill Creek.

Limonite in small quantities, under the forms of

compact brown oxide of iron, hematite, and bog ore, occurs in many places in this and Westchester county. The loose masses scattered over the surface of the earth seem to indicate important beds in Putnam. If surface indications are worthy of notice, a bed of hematite and brown iron ore will probably be found in the hills near the county poor-house.

Copper and Silver Ores.

Several mines have been opened in Putnam and Westchester counties, under the expectation of obtaining silver. I have examined a great number of ancient diggings in this county, where it is reported or imagined that silver has been, or is to be found ; but I have seen no indications worth pursuing, or any ore that is known with certainty to contain silver. These ores have not been analyzed, and it is not known that they even contain any silver, except from the common reports of the country that silver has been obtained from them. Almost all the diggings are in or contiguous to limestone. Many interesting mineral localities have been opened, and an abundance of crystallized minerals dug out, and prepared for the hand of the collector of these beautiful productions of nature.

It is, perhaps, superfluous to go into a detail of the numerous mining explorations in search of the precious metals in the Highlands ; suffice it to say, that superstition and the mineral rod have been freely employed, and credulous persons have permitted themselves to be imposed on, and in some instances have expended their all in explorations which any one versed in minerals, and acquainted with their associations, would have known from the beginning were hopelessly fruitless. Common pyrites and magnetic pyrites were repeatedly brought to me while I was stationed at the United States Military Academy as an instructor of chemistry, mineralogy, and geology, as specimens of gold ore, silver ore, and tin ore, by the mine hunters, or by those who had been imposed on. After examining mineral localities where lead and tin ores had been said to have been discovered, I have seen none in place, and have reason to believe that the specimens shown to me did not originate where they were said to have been found. A piece of metallic antimony was shown to me, and was said to have been found in this county; but it had the peculiar foliated, crystalline texture that is generally seen in that which has been melted, and which is different in aspect from the native antimony.

While on this subject, I will notice another fact that came under my observation. Coal was said to have been discovered in the primitive region of this county. I was shown a lump of beautiful *Mauch-Chunk Anthracite!!* which had been buried by some means unknown, and dug up;* and this was the reported coal mine. It is hoped that our citizens will no longer suffer themselves to be duped by designing persons into mining speculations, most of which have a baseless foundation.

Copper ore has been found in several places in this county and Westchester, but not in such quantities as to justify exploration. Pyritous copper and green carbonate of copper are found in small quantities in the gneiss rocks at Philips' mills, one mile and a quar-

* This was afterwards ascertained to have been buried by a designing person, with a view to get up an excitement, and or ganize a company to dig there for coal.

ter east of West Point; also at Philips' iron mine, eight miles north-east of Cold Spring landing.

Arsenical iron occurs in several places in this county, but the only locality known here, to which any practical importance is attached, is about four or five miles north-west from Carmel village, and about half a mile south-west of Pine Pond, in the township of Kent, near the serpentine marble quarry. This is one of the old mine holes from which silver is reported to have been obtained. The mine is now owned or leased by a mining company called the Hudson River Mining Company. It had been cleaned out when I saw it. The shaft is forty feet deep. Yellow, pulverulent sulphuret of arsenic covered the sides of the shaft and the timbers, wherever they had been covered by water, resulting from the decomposition of the arsenical sulphuret of iron. This latter mineral abounds there. It forms a bed or mass in hornblendic gneiss rock above the shaft, and is there undergoing decomposition, forming arseniate of iron. The ore does not, so far as I could perceive, form a vein, but is a mass; and from the surface indications, and from what I saw in the mine, there is a probability of the existence of a great quantity of this ore. The mine goes by the name of the *silver mine*, and it is stated that silver has been obtained from it, but the individual who is said to have analyzed it has no public name as a chemist; and until it shall be analyzed by a disinterested person, of reputation as an analytical chemist, confidence ought not to be reposed in the statement that it is a silver ore.

This kind of ore is wrought as a silver ore in Germany, where it contains some of the precious metal.

It is possible *this* may *also* contain it, and even should it be argentiferous, it may not contain enough of silver to make it worth separating. The ore contains much arsenic, and it may perhaps be profitably wrought to furnish the common white arsenic of the shops. It is well known that large quantities of this material are consumed for various purposes in this country, such as the manufacture of shot, flint glass, medicinal preparations, &c., and the supply is at present derived from Germany. This mine would probably supply the demands of commerce.

Titanium ore has been found in several places in this county. At almost every locality where augite and scapolite are found (and the localities are numerous), sphene or the silico-calcareous oxide of titanium is also found associated. Sphene, beautifully crystallized, was discovered by Dr. Barratt at Cold Spring landing, in 1822, during the excavations for the foundation of the long block of buildings next the shore on the north side of the village.* Specimens were obtained at that place in abundance by Dr. Barratt, and more beautiful than any that I have seen from any other part of the country. Titanium has, however, been applied to but one useful purpose, and that of comparatively trifling importance, viz., for tinging the enamel of artificial teeth of a slight yellowish color, like the natural teeth. It has also been found in a great number of localities in the adjoining county of Orange. Wherever observed, it is associated with augite, scapolite, and limestone. It seems almost confined in the first district to those rocks we

* The long range of two-story old buildings belonging to Mr. Philips, and known by the name of the "*Barracks.*"

have described as metamorphic. An ore of *cerium*, called allanite, occurs in several localities in Philipstown, within two miles of Fort Montgomery, and it is thought other ores of this metal were observed some years ago. One of them was partially examined, and re-agents showed some of the traits of that metal.

Crystallized Serpentine.—Serpentine was found crystallized at Huestis's quarry in Philipstown, first by Dr. Barratt in 1821, and secondly by Cadet Ward, November 5th, 1831. By means of crystals from this locality I have been enabled to ascertain the primary form and its elements. November 5th, 1831, a specimen of serpentine was handed me for examination, by Cadet Ward, from Huestis's quarry. I immediately observed traces of crystallization, there being numerous well characterized laminæ, and showing tolerably brilliant cleavage planes also in other directions. Much of the serpentine in some parts of the hill-side at Huestis's quarry is granular and some is laminated. Cadet Bailey, now Professor Bailey, at West Point, also presented me with a specimen of crystallized serpentine imbedded in carbonate of lime. There were several hexagonal crystals terminated, one of which was nearly perfect. The crystals and crystalline masses are more perfect than any I have seen from the eastern locality.

Primary Rocks.—The rocks of this county are numerous, many of them are applied to useful purposes, and they are everywhere abundant, and are seen cropping out from the surface of almost every hill and ravine. The same kind of rocks are found in New York, Westchester, and Dutchess counties. The principal rocks are, 1. granite ; 2. sienite ; 3. gneiss ; 4. mica slate ; 5. augite rock ; 6. green stone and hornblende rocks; 7. quartz rocks ; 8. talcose slate ; 9. limestone ; 10. serpentine ; 11. steatite.

The five latter rocks have already been described as metamorphic rocks.

1. *Granite.*—This rock occurs abundantly in Putnam, Dutchess, Westchester, and New York counties. It presents all varieties of texture, from a very coarse grained rock, to one almost perfectly compact. In color it varies as much as in texture. It is white, red, grey, yellowish, and bluish-grey, according to the color of the minerals forming it. The color of the felspar usually determines that of the mass. It occurs in beds, in veins, in interstratified masses, and in knots, knolls, and protruding masses, in which no connexion with veins or beds have been traced. The more common mode of its occurrence is in beds ten to one hundred feet thick, interstratified with gneiss. Some of the granite is too coarse for use as a building material. Some is too compact and hard, being, in fact, *erutie ;* others are well adapted for building. Different localities show a great variety in strength, and in the ease or difficulty of dressing, as well as in the ease of quarrying and the magnitude of the blocks that can be procured. In the Geological Report of 1838, it was mentioned that many places would undoubtedly be found in the Highlands, where fine quarries would be opened, and furnish " building materials of the best quality, and which would endure the changes of our variable climate for ages without decay or disintegration."

The investigations subsequent to that time have

verified the prediction that such localities might be found. The materials are of the best quality, easily quarried in large blocks, suitable for columns, cornices, &c., easily dressed, enduring as time, as the naked crags themselves will testify; and several of the localities, which were unknown to their owners, are so convenient to water transport that the blocks can be swung directly on board vessels in the Hudson, by means of cranes. When we consider the value attached to the quarries in Maine, Massachusetts, and Connecticut, where, in most places, it is necessary to haul the stone, either on a common road, or construct a railway to navigable water, a distance from half a mile to six or seven miles,—and observe that notwithstanding all these disadvantages, the great outlay of capital, and the distance to the market, they make it a profitable business,—we may begin to appreciate the importance of having inexhaustible quantities of materials, as good, as beautiful, as durable, and as easily quarried and dressed, on our own waters, within forty or sixty miles of the city of New York, and so convenient to shipment that no railroads and hauling are required..

Blunt's Quarry.—This is located on the south side of Breakneck point, near the line between Putnam and Dutchess counties. There is an inexhaustible supply of a material of the best quality. It is a bluish gray granitic rock, composed mostly of a dark-colored felspar, with some hornblende, quartz, and occasionally a little mica. It is more properly sienite than granite. It is scarcely as dark as the Quincy granite or sienite, while it is as beautiful, has about the same strength, splits as well, and is as easily dressed. The stone from this quarry has been extensively used in the construction of the Delaware Breakwater, of Fort Calhoun, and Fortress Monroe. The mountain rock has not been quarried at this place, but only the large masses that have tumbled from the cliffs above. It is not possible to give an accurate estimate of the quantity of granite in this vicinity, but there may be in the end of the mountain five hundred acres, with an average depth of five hundred feet, or 803,640 cubic yards to the acre, or 401,720,000 cubic yards on five hundred acres.

Blunt's quarry is located on the immediate shore of the Hudson river, but on account of the flats, the stone has been hauled about sixty rods to a landing. This quarry bids fair to become valuable ; but there is one disadvantage that *may perhaps* operate as a drawback to its advantageous position. It is overhung by a precipice of several hundred feet in height ; and in the quarrying operations, the heavy blasts may bring down hundreds of thousands of tons of rock which can be useful only for dock stone and ballast.

Highland Granite Company's Quarry.—This quarry is principally owned by Messrs. Howard and Holdane. It is located one-fourth of a mile from the Hudson river, and half-a-mile east of Blunt's quarry near Breakneck point, and about two miles from Cold Spring. It is elevated about four hundred feet above the Hudson, in full view of the river. The stone is of excellent quality, and splits easily into large blocks. It is composed principally of felspar, with a little hornblende, and is indistinctly stratified ; or at least it lies in thick heavy beds, with parallel seams six to twelve feet apart, and which are slightly inclined to the horizon. The quarry is inexhaustible, and ought to be very valuable. This quarry is on part of the bed of granitic rock described under Blunt's quarry. The stone is now hauled to the landing, about one-fourth of a mile, at an expense of three cents per cubic foot, or forty-two cents per ton. Much of it is sent to Sing Sing, for the culverts and aqueduct bridges ; and the freight to that place is four cents per foot, or fifty-six cents per ton. It is delivered at Sing Sing in blocks of ten cubic feet and over, at thirty-five cents per cubic foot, or five dollars and ninety cents per ton. The dressing of this stone for the arches, is done at fourteen and a half cents per superficial foot ; and about two and a half superficial feet are dressed to the cubic foot, which make the stone dressed, ready for the arches, cost seventy cents per cubic foot, or nine dollars and ninety-seven and a half cents per ton.

This quarry is capable of being worked at least seventy yards in depth, over an area of several acres ; and allowing a profit of one dollar per cubic yard, which is a low estimate, and 4840 square yards to the acre, fifty yards in depth ought, in the course of working, to give a profit of 242,000 dollars to the acre.

Stony Point, one half of a mile north-west of Cold Spring.—This is a rocky peninsula, stretching into the Hudson about one-fourth of a mile. It is composed of gneissoid rocks, except the north-west point of the peninsula, which is a granitic rock of the same character as that of Blunt's and the Highland Company's quarries. About two acres of this peninsula are covered by this rock, to an estimated mean depth of forty-five feet above high water mark ; and it may be estimated that there are 145,200 cubic yards of granite capable of exploration on this point. It may apparently be split out in masses of any size, up to one hundred tons or more, in regular blocks; and it lies immediately on the Hudson river, and with such a depth of water that large vessels may come immediately alongside of the rocks to be quarried, so that the blocks may be swung on board with a crane. Stonypoint is owned by Mr. Philips* of Philipstown, who was not aware of the existence of such a location for a granite quarry, until he was informed of it during the progress of the survey of Putnam County in 1840.

Philips's Quarry.—This belongs to the same gentleman as the preceding. It is located on the Philips estate, about half-a-mile from the Hudson river, and one and a half miles east-north-east of West Point. The rock is perfectly indestructible, and would be called granite by those who should see the blocks without seeing the quarry. It is gneiss, in thick layers or plates, which have a slight inclination to the west, while the grain of the rock is nearly vertical. It splits easily, both in the direction of the grain and across it. It may be procured in the form of blocks of five to ten or more feet square, and of the thickness of the plates of rock, which are from one to four feet thick. Some masses were seen which had been split off for columns for store fronts, twelve to fourteen feet long, by one and a-half, one and three-fourths, and two feet square.

The rock at this quarry is of a light grey color, almost white, and is a beautiful material for building. It is durable, of sufficient strength, easily dressed and

* Now owned by Anderson & Co.

easily quarried, and the stone can be transported to the banks of the Hudson for three to four cents per cubic foot.

The extent of this rock was not ascertained; but there is an area of at least ten acres, with a mean depth of sixty feet, or 26,136,000 cubic feet, or 968,000 cubic yards of this granitic gneiss, or about 1,900,000 tons.

There is a location suitable for quarrying in this county, about three and a half miles below West Point, and near the Cotton rock. The granite or granitic gneiss is of good quality, of a light grey color, and durable. This locality was not examined closely; but from the general aspect of the rock, it is believed to be a good location for a quarry. Beautiful light grey granite was seen in abundance from one to two and a half miles north-west of Boyd's corners. It is as durable as time, and may be procured in any quantity; but its distance from easy transportation by water or railroad will prevent its use at present beyond the neighborhood.

It is estimated that several millions of dollars are annually paid out of the city of New York, and the towns on the Hudson river, for building stone brought from beyond the limits of the State; while we have within our own boundaries, and near the markets, inexhaustible supplies of equally good quality, which can be quarried, shipped, and hauled at less expense than the stone we now import from Maine, New-Hampshire, Massachusetts, and Connecticut. The granites of the Hudson river *must*, then, soon be wrought and sent to market, and the quarries will become very valuable.

GEOLOGY. 57

2. *Sienite.*—This rock abounds in some parts of Putnam and Westchester counties. In Westchester county, it approaches in its characters to the "Quincy granite" of Massachusetts, and would probably make as beautiful and durable a material for building as that which is so justly celebrated. In *this* county the sienite is generally coarse-grained, of a reddish color, spotted with black crystalline and irregular masses of hornblende. This rock passes into hornblende slate and hornblende gneiss on the one hand, and into hornblende rock on the other. No localities were seen in this county where this rock would be available for economical uses, except the granitic sienite, which has already been mentioned under the head of granite, as occurring in Breakneck mountain, and at Stony point above Cold Spring.*

The sienite rock of the *Highlands* is of two kinds. One is a coarse granitic, aggregate of white or reddish felspar and black hornblende, sometimes also containing epidote and grains of magnetic oxide of iron, like that at the base of Bull-hill, one and a half miles north-north-west of Cold Spring village, on the shore of the Hudson; and at the Target rock, on Constitution island, opposite West Point; the other is composed

* The mountain at the north-west corner of Putnam county, is frequently called *Anthony's Nose* and *Anthony's Face*, in consequence of the profile bearing a rude resemblance to the human face, that may be seen in one position in passing it; but *Breakneck mountain* is the name by which it is generally known. Anthony's Nose mountain is at the southwest corner of Putnam county, opposite Fort Montgomery. Stony point, above Cold Spring, I propose to call *Quarry point*, to distinguish it from Stony point in Rockland county, a place of much notoriety in the annals of the Revolution.

mostly of felspar of a dark greenish or sometimes yellowish and brownish color, with some quartz and hornblende. The latter is black or green, and sometimes passes into that described under the name of hornblende rock, where the hornblende is arranged in stripes through the rock. The felspar in this kind of sienite is occasionally opalescent, but is distinct in characters from that from the north part of the State, and which is seen in boulders and blocks on the slopes of the mountains in the Highlands.

3. *Gneiss.*—Gneiss is the predominant rock in Putnam, New York, and Westchester counties. It varies greatly in external aspect and in composition, in different parts of the tract under investigation. Its color is dependent upon the relative abundance of its constituents, which are variously colored in different localities. The felspar is white, reddish, or of a bluish grey; the mica is black, brown, yellow, copper-colored, and white; the quartz is white, grey, or smoky. In some places mica abounds in the rock, and it approaches to mica slate, but more commonly the felspar is most abundant, and gives character to the rock.

Much of the gneiss in the Highlands of the counties under consideration is a hornblendic gneiss, in which the mica is wholly or in part replaced by hornblende.

A range of granitic gneiss, of a light color, passes through Putnam and a part of Westchester county. It extends through Carmel, near Pine pond, by Mahopack pond; thence southwardly, and crosses the turnpike from Peekskill to Danbury. Another bed extends from Boyd's corners, and crosses the Peekskill and Danbury turnpike about five or six miles from the former place. These beds are quarried, to a small extent, for use in the vicinity; but they are too remote from water transport, for quarrying at present for a more distant market. It is durable, of a light grey color, easily split from the quarry, and easily dressed. If these strata reach the Hudson river, they are believed to have changed so much in aspect and quality in building stone, as not to have been recognized as the same beds.

4. *Mica Slate.*—This rock has a very limited distribution in Putnam county. Where it does occur, it seems to be a modification of gneiss, the mica becoming predominant, while within a short distance the rock resumes its characters of gneiss. No locality was observed where there is a prospect of valuable quarries of flagging stone of this kind of rock being opened, near water transport.

5. *Augite Rock.*—This rock occurs in a great number of localities in Putnam county, and in a few in Westchester county. It is sometimes intermixed with felspar, but more commonly it is either by itself, or mixed with the various minerals that are usually associated with it. It occurs at most of the celebrated mineral localities in the Highlands. It is of all shades of color, from white through grey and green of various shades to black, and from compact through various grades of granular to broad foliated masses, in the forms of fassaite, coccolite, common augite, sahlite, crystallized augite, and diopside. This rock has not been applied to any useful purpose.

It is believed that this rock might, with propriety, have been described among the Metamorphic rocks. It has rarely been found except in connection with such rocks, and is almost constantly with scapolite, granular limestone, and hornblende. It generally also, in Putnam county, has magnesian garnet and plumbago associated. This rock forms extensive masses between Anthony's Nose and Sugarloaf mountains, along the eastern side of the Hudson, between the shore and the base of the mountains. Between the "Old silver mine" and the Hudson, about four or five miles south of West Point, it contains large quantities of crystallized, massive, and granular magnesian garnet. The augite is dark green, and sometimes black, containing plates and hexagonal scales of plumbago. This rock scarcely corresponds with augite rock as described in systems of geology, as it does not generally contain *felspar*. Most frequently it is an aggregate of augite and scapolite, augite and carbonate of lime, or augite and magnesian garnet, and sometimes augite and mica.

A locality of white or rather grey augite, may be examined on the shore of the Hudson, about opposite Buttermilk-falls, and two miles south-east of West Point, a little above the point from which Arnold escaped. The augite here forms a heavy bed in gneiss rock associated with limestone. The augite is crystalline, grey, and contains scales and hexagonal plates of plumbago. A vein of mineral that I suppose to be the pyrophyllite, traverses the augite rock. The mineral from this locality has the aspect of silvery mica, which can be dug out at the vein in masses so as to give plates of two or three inches in diameter, in rhombic and hexagonal crystals like mica; but the plates have not so much elasticity as mica, nor so much unctuosity as talc. The plates of this mineral, when heated, exfoliate, and spread out like the vermiculite of Rhode Island, so that a plate of an eighth of an inch thick before being heated, becomes one-half to one and a half inches thick in the fire, the laminæ all separating, but remaining still attached to each other.

Another interesting locality of augite rock is on Anthony's Nose mountain, at the " White mine." It is here associated with a bed of brown spar, containing magnetic oxide of iron and plumbago. An ore of *cerium* is supposed to have been observed at this place. Rocks of augite containing scapolite and sphene, were seen in many places on the shore of the Hudson, at the southern base of Anthony's Nose mountain, but the localities from which they had fallen in the cliffs above were not traced out. Augite containing sphene, scapolite, and associated with verd-antique, diallage and hornblende, occurs at the base of the cliffs of Bull-hill, near the shore of the Hudson; but the beds from which they had fallen, although some explorations were made, were not seen. The augite occurs under various forms, as green, yellowish, grey, crystalline, crystallized granular (coccolite of white, green, grey, yellowish, and red), fibrous, and in acicular crystals.

Augite also occurs at Huestis's quarry in Philipstown, as augite, white and green coccolite, diopside, and sahlite.

Cold Spring was an interesting locality of augite rock some years ago, but a block of buildings has been raised over the locality where so many beautiful specimens were procured. The augite rock is there associated with gneiss and granite, and contains scapolite and sphene in abundance. The largest and most beautiful crystals of sphene I have ever seen were obtained at this place by Dr. Barratt in 1822, now of Middletown, Connecticut. Augite is so common a rock in Putnam county, that it is unnecessary to multiply localities.

6. *Greenstone.*—This rock traverses the strata in many places in Putnam and Westchester counties. In some places it has the aspect of compact trap, like basalt, but more frequently the hornblende predominates and gives its characters to the rock. It traverses, and is intertruded in sheets and irregular masses among the gneiss and other rocks, in the same way as granite and sienite; and many of the masses classed with this rock would be classed with sienite, but for the fineness of the grain, being of about the texture of a sandstone, composed of black hornblende with grains of white and grey felspar.

Well characterized dykes of greenstone of the basaltic kind were seen in a few places in *this* county. One was near the mills* north-east of Huestis's quarry; and another near the road from Cold Spring to Davenport's corners, about two and a half miles from the former place. The speckled greenstone in which hornblende prevails, may be seen abundantly in almost every part of the Highlands of Putnam and Orange counties.

Hornblendic Rocks.—This is a convenient repository for those rocks that are not so perfectly characterized as to be included under the preceding heads. Hornblendic rocks form a very considerable proportion of the mass of the Highlands in Putnam, and in

* Knapp's Mills.

fact in Rockland and Orange counties; but those parts composed of sienite, hornblendic gneiss, hornblende slate, and greenstone, have been described. Perhaps the remainder classed under this head might properly have been described as greenstone, for they have the geological relations of that rock, being evidently in many instances an intrusive rock; but very frequently it is almost pure hornblende, and could not, in conformity with the generally received composition of greenstone, be described as such.

Hornblende rock is abundant in Anthony's Nose mountain, between Anthony's Nose and Royahook. Hornblende forms a constituent of a large share of the rocks of this mountain.

Hornblende is also common between Anthony's Nose and Sugarloaf mountains. Greenstone, hornblendic gneiss, and hornblende rock occur at the northern base of the hill at the laumonite locality,* about one hundred to two hundred yards below Philips's mill, one and a quarter miles east-north-east of West Point.

The hornblende rock is common on Bull-hill, the mountain north of Cold Spring.

The hornblendic rocks are constantly associated with the beds of magnetic oxide of iron, which are so numerous in the Highlands.

* This locality of laumonite and stilbite has been said to be exhausted. It is not. It is a vein of decomposing felspar, two and a half to four feet wide, in which the laumonite and stilbite crystals abound. I had a blast put in the vein in 1829, and obtained an abundance of specimens, showing these small but perfect crystals, in groups, in the cavities of the felspar. Many wagon-loads could probably be obtained. Much of the felspar is dark-colored glassy felspar.

On the turnpike from Cold Spring to Carmel, the rocks are gneiss and micaceous gneiss, hornblendic gneiss with beds and veins of granite, greenstone, and hornblende rock. The gneiss on the eastern declivity of the mountain, for some distance from the crest, is hornblendic, and the dip is to the eastward, where it is not vertical.

The heavy swell of land, composed in part of talcy slate, east of the limestone that was seen near Bunnell's forge, is bounded on the east by Horton's pond and its outlet. Gneiss was frequently seen in place after passing into the valley of the pond; and on the eastern side of its outlet, the rock had much the aspect and composition of some of the felspathic and sienitic rocks south-east of Peekskill, though they had more of a granitic aspect. The rocks at *Cold Spring landing* are gneiss, hornblendic gneiss, and granite. The strata have a north-north-east and south-south-west direction, and the dip is vertical at the south point.

Constitution Island, between Cold Spring and West Point, is composed of gneiss, hornblendic gneiss, granite, and sienite. The stratification is much confused, and some of the rocks have a strike transverse to the usual direction, viz., north-west and south-east. This appears to be on the transverse line of disturbance that has been observed farther east-south-east in several places, and on the west-north-west near the cascade, and on the mountain farther west. Granite and sienite form the Target rock, a high cliff on the south-west side of the island, and granite forms some of the points farther north. Hornblendic rocks (gneissoid) form the shore a little north of the Target rock; they lie in strata dipping at high angles to the north-east, and some are nearly vertical.

Flat *rock* at Mr. Arden's boat landing, two and a half miles south of West Point, is granite. The geological explorer can scarcely fail of finding numerous localities of granite, gneiss, sienite, greenstone, hornblende rock, augite, limestone, etc., in exploring the shores of the Hudson through the Highlands.

ORES OF THE HIGHLANDS.

Magnetic Oxide of Iron.

This ore is confined to the highlands, and abounds in Putnam county. Several mines are already wrought, and many more are capable of exploration. They form masses in gneiss and hornblendic gneiss rocks, which by casual examination would be called beds; but after a careful investigation of the facts, I think they may be called veins. Their course is parallel to the line of bearing of the strata, and they lie parallel to the layers of the rock; but by close examination, it is found that in several instances, after continuing with this parallelism for a certain distance, the ore crosses a stratum of rock, and then resumes its parallelism; then crosses obliquely another, and so on. In other places, where a great bed of the ore occurs at some depth, only a few small stripes of ore penetrate through the superincumbent mass to the surface, as if the rocks had been cracked asunder, and these small seams of ore had been forced up from the main mass below.

The beds of veins of magnetic iron ore lie either vertical, or dipping to the east-south-east, at an angle corresponding nearly to the dip of the strata. One example only was observed where its dip was to the west-north-west, viz., at the Stewart mine. The ore is very variable in quality. In some it is nearly pure magnetic oxide of iron; in others, it is intermixed more or less with the materials of the contiguous rocks; in others, it is mingled with pyrites and with other minerals. Two main veins of this ore will be described under the names of the Philips's vein and the Simewog vein. Numerous localities are known where this ore occurs, and where it has long been dug in small quantities. They will be mentioned under the head of local details.

LOCAL DETAILS.

A bed of magnetic oxide of iron has been opened on Breakneck mountain, and several tons taken from it. The extent of the bed is not known, and the ore has not, it is believed, been smelted.

Another bed has been opened on the north-east part of Constitution Island, opposite the West Point Foundry. Another was opened in the middle of the island. The ore occurs disseminated in granite near the redoubt, above the Target rock on Constitution Island. Magnetic oxide of iron is thickly disseminated in limestone, near Philips's mill, one and a quarter miles east of West Point; and it is found in that stratum of limestone in many places, from the above locality to near half a mile south of the "Cotton rock," to a distance of three miles.

It also occurs in the granite rock that is associated with augite and limestone rocks near the "Old silver mine," three-quarters of a mile south-east of Conshook island, and one mile north-east of Anthony's Nose mountain.

A bed was opened many years ago on Anthony's Nose mountain, but it contained much pyrites and crystallized phosphate of lime, both of which injure the ore for the manufacture of iron.

The brown spar at the "White mine," about one mile east of the western summit of Anthony's Nose, contains magnetic oxide of iron disseminated. A locality of magnetic oxide of iron occurs on Mr. Tilly Foster's farm, two and a half miles south-east from Putnam Court-house.

The ore forms a large part of a hill about one hundred yards long, ten to forty feet broad, and elevated twenty to thirty feet above the ground adjoining. Some hundreds, perhaps thousands of tons of ore can be easily procured at this place, without digging below the level of the hill. It is associated with serpentine, with limestone containing brucite or boltonite, and with green mica. The mass of ore is bounded by gneiss on the east; and serpentine, limestone, and verd-antique seem to form its western boundary. It was thought that some *chromated oxide of iron* was observed here, but no examination has been made to ascertain that point. Another ore bed was discovered some years ago about half a mile south-west of the preceding, on land belonging to the Misses Fowler. Some tons were dug out, but I do not know whether any has been smelted. The ore is here mixed with manganesian garnet, augite, and hornblende.

The *Simewog vein* passes through Simewog hill, and was traced one and a half miles south-south-west on Mr. Jedediah Wood's farm; and it is supposed to continue still farther south-south-west, as ore has been dug in that direction about one mile south-south-west from Mr. Wood's house. This vein was formerly extensively worked at Simewog hill, and the mine is called Townsend's mine.

This mine was the first known and first worked in this part of the country. The ore was carted to great distances, and shipped on the North river, to some of the towns on Long Island sound, and various parts of the country. The largest portion of the ore was carried to Danbury in Connecticut, and was there an article of traffic. It has not been wrought for twenty or thirty years, in consequence of other beds having been found in more convenient locations for smelting and transport. Fifty thousand tons of ore, at least, have been taken from this mine, estimating four tons to the cubic yard; and one hundred thousand tons more may probably be taken from the vein in Simewog hill, without going below the level of the small stream which flows across the ore bed. Should it ever be necessary to obtain this ore in quantity (as is probable, from the prospect of the New-York and Albany railroad passing up the valley on the east side of the hill), at least one million tons may be calculated on, above the water-level of the Croton river, which flows along the base of the hill, and free from the expense of drainage, by driving an adit level from the level of the Croton, a distance of three hundred or four hundred yards to intersect the vein. This vein of ore has also been worked to the extent of several thousand tons, near the road and north of the little stream mentioned above crossing the vein. The vein here is from eight to fourteen feet thick, and nearly vertical in position, between strata of gneiss and hornblendic gneiss which dip seventy to eighty-five degrees to the east-south-east. On Simewog hill, one-fourth of a mile south, the vein is from three to twenty feet thick, associated with similar rocks and with granite. It has been wrought on Simewog hill from thirty to sixty feet or more in depth, over a length of three hundred to four hundred yards. It is scarcely doubted, from the observations made, that this vein is at least two miles in length, with an average width of six feet. Its depth cannot be estimated, but it is presumed that the labor of ages could not exhaust it in depth, as the bottoms of such veins have never, in any country, been found.

In the estimates above, the calculation is based upon the vein being wrought down to the water-level of the adjacent valley.

This ore bed seems to be a vein, although its strike is the same as that of the strata. In the excavations on Simewog or Mine hill, the bed or vein seems to have crossed the strata very irregularly and obliquely, and similar to the vein in the Shawangunk mountain at the Sullivan mine, running between the strata for a certain distance, then crossing obliquely between two other strata, and so on.

The Phillips' vein has been traced at short intervals for about eight miles, and is presumed to be continuous through this distance, except where it is interrupted by dykes and transverse heaves of the strata. Many mines have been opened on this vein, and several of them are now worked.

The Cold Spring and Patterson turnpike crosses this vein of iron ore near the crest of the mountain, about nine miles from Cold Spring landing. There is an opening near the road, and near this crossing, where some ore has been dug. Here the ore seems injected in little sheets, veins, and beds, through the gneiss rock, so as to form one-fourth to three-fourths of its mass through a horizontal thickness (as the strata are vertical) of thirty to thirty-five feet. Pyrites abound in a portion of the bed. The ore is easily traced along its course, as it shows itself distinctly along the line of bearing of the strata, disseminated, and forming black stripes in the rock. Near the house, one or two hundred yards farther south-south-west, another small opening has been made. One hundred to two hundred yards farther south-south-west on the line of the vein, a larger excavation has been made, and five hundred to eight hundred tons of the ore thrown out; but it is here so much intermixed with pyrites as to be unfit for smelting, until the pyrites shall have decomposed. Some hundred yards farther south-south-west on the line of the vein, another opening has been made next the marsh, and it is continued down the hill. The ore is here more or less intermixed with the rock, with a breadth of ten to twenty feet, and the gneiss and hornblendic gneiss rocks associated dip to the east-south-east at an angle of about sixty degrees. Farther down the hill are two main openings, which go by the name of Phillips's mine. The ore in some parts of the upper mine is more or less intermixed with copper pyrites, which injures the quality of the iron. The mine has been wrought badly, timbers being used to prop the overhanging rock, and great masses have crushed in and filled most of the mine.

The lower mine, where the whim is placed, has a

solid rock roof, a part of the ore bed having been left in the top of the hill, while the mine has been worked below. The ore bed is here fifteen to twenty feet wide, and has been wrought thirty to forty feet in depth, over a length of fifty yards, This mine is not worked open to the day like a quarry, but a drift crosses the strata to the mass of ore, and it is worked at and below this level, along the course of the vein under a cover of rock. The ore does not show itself very distinctly in the over-lying rock. The ore here is nearly a pure magnetic oxide of iron, and twenty thousand to thirty thousand tons have probably been taken from these two mines.

Other openings have been made along the line of the vein for about a half a mile farther to the south-south-west, and some three thousand to five thousand tons of ore probably removed. The rock in which this part of the vein thus far described is contained, is mostly felspar, with some bluish quartz; hornblende is also common. The felspar is sometimes pearly in lustre and gray in color, with wrinkled and bent faces, as if it had been soft, and subjected to forces acting in different directions.

Other openings along the course of this vein were traced for half or three-quarters of a mile in a south-westerly direction. Hornblende abounds in the rocks associated with the iron ore.

The next mine that is worked to any extent on this vein, is the Stewart mine. It is about twelve feet thick of pure ore, and four feet more of lean ore. The former is much used in forges, the latter in the blast furnace. The ore at this mine is purer than that of any other mine I have seen, and is easily worked in the forge. It is granular, and easily broken and crumbled into grains about the size of BB shot, and is called by the miners "shot ore." The vein lies between strata of felspathic gneiss, which dip to the west-north-west about seventy degrees. This mine is on the east side of the mountain crest, and about one hundred to two hundred feet above a marsh, with a steep declivity, and might easily be wrought to that depth without drainage, by driving an adit level to intersect the vein.

About half a mile south-south-west is another opening by the road-side, where some ore has been dug; but it is lean, and much intermixed with the gneiss rock. About three-fourths of a mile south-south-west of this is the Denny mine. It is about two and a half miles east-north-east of Warren's tavern,* in Philipstown, in a straight line on one of the crests of the eastern ridge of the Highlands. The ore seems to have been injected among the rocks. In some places it forms regular stripes on the surface of the rock, parallel to the line of bearing; in others, there are scarcely any indications on the surface, while extensive masses exist a short distance below. This cap of rock over the ore is frequently called by the miners a *rider*, and the ore below, the *horse*. The mine now at work north of the house, is about thirty feet deep, and the vein of solid ore twenty-five feet wide, overlaid by a cap or rider of rock which contains but little ore. Most of the ore is very compact and pure, but some contains hornblende. Much of the felspathic rock contiguous to the vein is injected with thin veins of ore from one-eighth to one inch thick. Two hundred yards south-south-west is another opening, from which

---

* Now owned by Justis Nelson, Esq.

much ore has been taken. This place has been excavated to a depth of sixty feet, and the vein is twenty to thirty feet wide. Twenty thousand to thirty thousand tons of ore at least have been removed. Contiguous to this opening is another, thirty feet deep to the water, with a sheet of rock five or six feet thick, between two divisions of the vein. The rocks on each side of the vein are more or less injected with thin veins of ore. From examining the locality, many suppose that the ore has been injected into the cracks and crevices of the rock when broken up by some upheave.

This ore is deliverable at the Cold Spring furnace, and at the wharf at Cold Spring, for three dollars per ton; and mined as it is, scarcely any profit can be realized at this price. The quantity mined here is six hundred tons per annum.*

The Coalgrove mine is about one or one and a half miles south-south-west of the Denny mine; it is gneiss. The vein is narrow at the surface, but at the depth of twelve feet it is four feet wide. The ore is of an excellent quality, very rich, and well adapted for the forge, and will undoubtedly make an excellent

---

* Here holes are dug down in this ore bed or immense vein of iron ore, and water accumulates unless pumped out, or drawn out by a tub and whim. By the present mode of mining, two men, a boy and horse, are required to tend the whim for drawing up water and ore; when if properly worked, the same quantity of ore could easily be wheeled out by one man or boy, or carts could enter the mine and load, and dispense with this kind of labor entirely. A small and short adit level from the hillside east of the mine, would lay the ore bed dry for a hundred feet or more in depth for a considerable distance.

iron. The distance from this mine to the furnace and Cold Spring landing, is less than from the other mines.*

The Gouverneur mine is about one and a half miles south-south-west of the Coalgrove mine, and four miles east of the Philips's manor house, at the southeast corner of the "water lot." The ore is much intermixed in the rock, but would perhaps work well, mixed with other ores, to flux out the felspar and other minerals. It may probably be purer farther down. It has been opened in several places along the crest of the mountain to a depth from three to twelve feet. The ore is disseminated in the gneiss and granitic rock, through a thickness of five to twenty feet. The strata are nearly vertical. It is on one of the crests of the eastern ridge of the Highlands, west of Peekskill hollow. A slight opening has been made about three-fourths of a mile north-north-east of the Gouverneur mine, between that and the Coalgrove mine. The ore is titaniferous, and in lumps, and disseminated in the rock. The vein is six to twelve feet wide. It may perhaps be worked by picking the ore, so as to separate the lumps from the gangue.

The mines and openings just described are the principal ones on the Philips vein, but the ore can be found along almost the whole line.

It follows the crest of the east ridge of the Highlands a distance of at least eight miles. The breadth of this vein has been mentioned at different places from three to thirty feet wide; its average is probably about twelve feet, and its length, as now known, about fourteen thousand yards. If the mean average

* The Kemble mine is a short distance north-north-east of the Coalgrove mine, and on Philips's vein.

of the vein be supposed to be half its bulk of ore, every cubic yard will contain about two tons of ore, and would yield at least one ton of iron, or each yard in depth would make fifty-six thousand tons of iron. The vein, by proper working, can be mined to a mean depth of one hundred yards, without expense of drainage more than the proper opening of adits. We may place the workable produce of this vein, above the water level of the adjacent valleys, at 5,600,000 tons of iron. The phenomena of the mines in many places on this vein induce the idea of igneous injection, connected with a powerful up-heaving force. The felspar is often pearly, wrinkled, and with bent laminæ. The appearance of hyalite, a mineral usually associated with volcanic and trap rocks; the apparent injection in veins among the seams and crevices of the rock; the appearance of the softening of the gneiss and bending its layers like a flowing slag seem to point to an igneous origin of this vein. It often has the appearance of a bed, and at other times of a vein ramifying from a main mass between the strata, and at other times cutting obliquely across them, but still having its out crop parallel to the line of bearing.

The Cold Spring furnace* is the only blast furnace in operation in the counties of New York, Westchester, and Putnam. It is supplied with magnetic oxide of iron from the Philips mine, the Denny mine in Putnam county, and the Townsend mine in Canterbury, and the O'Niel mine in Warwick, Orange county. These ores are mixed in certain proportions, and flux each other easily with a small addition of the

* Discontinued.

Sing Sing limestone. The produce of this furnace is from one thousand to fourteen hundred tons of pig iron per annum. Bunnell's forge* in Philipstown is believed to be the only one in operation in the counties under consideration. It is supplied with the shot ore of the Stewart mine.

Localities of Peat and Marl.—About thirty acres of this alluvion is found near the east side of Lake Mahopack in the town of Carmel; five hundred acres near Patterson; eight acres two miles east-north-east of West Point, in Philipstown; twenty acres near the head of the *Sunk* lot, eight miles from Cold Spring, on the road to Carmel village; twenty acres on the road from Carmel village to Patterson; six acres four miles south-east of Pecksville; twenty acres in Philipstown, east of Stewart's iron mine; ten acres in Philipstown, half a mile south of the last-mentioned locality; fifty acres in Philipstown, near Davenport's corners, five miles north-east of Cold Spring; and twenty acres in Philipstown, in the south-east part, near the Hon. Abraham and Saxton Smith's. Peat is probably abundant in the meadows near Constitution Island, though it has not been particularly examined.

The mud-flats near Constitution Island, are all increasing slowly, and from a variety of causes, such as vegetable decompositions, the silt and mud deposited from the water, and the growth and decay of molluscous and other animals. They have increased more rapidly during the last twenty years than before, in consequence of the greater amount of cultivated land causing a greater amount of earthy materials to

* Discontinued.

be transported by the rains and surface waters into the Hudson. These flats will eventually become meadows, but the time may be far distant. The flats along the right bank of the Hudson, opposite West Point, both below Gee's point and near Camptown, have grown sensibly more shallow within the last fifteen years. The same may be said of the flats between Constitution island and Gouverneur's landing, opposite West Point, and between Constitution Island and Cold Spring.*

THE PATENT.

This county was patented in 1697, by Adolph Philips, a merchant then residing in the city of New York. As shown by the Patent, it included "Pollepells Island," and contained more land than is now embraced by this county. Adolph Philips, or *Philipse*, as it was formerly written, was a bachelor, and an uncle to Capt. Frederick Philips, deceased; and great uncle to Mrs. Mary Gouverneur, of Highland Grange, Philipstown. Previous to this time, a brother or cousin to Adolph received a patent for a tract of land

---

* In 1822, sloops used to come in at the Foundry dock, about half-way between Cold Spring and the West Point foundry, to take in their freight of cannon and other castings; but the water has become so shoal that for some years past it has not been possible, and they now load at Cold Spring.

in Westchester county, "originally comprising not less than 20 miles square, bounded West by the Hudson, and lying south of the mouth of the Croton." This patent was granted to Frederick Philips, in 1680. We at first made an attempt to trace back the genealogy of this family; but, having been informed that an analysis of that matter would be given shortly in a work on Westchester county, we ceased our inquiry, expecting to see it more fully portrayed than our means of information would enable us to give. Besides, it was more proper that it should appear in a work on Westchester, as the elder branch of the family first settled there.

This patent in Dutchess County, covering nearly the whole of the Highlands, was inherited by the father of Capt. Frederick Philips, who left three children, one son and two daughters, viz.: Frederick; Mary, who married Roger Morris, a major in the British army; and the wife of Col. Beverly Robinson, also of that army, whose Christian name is unknown to us. The land embraced by this patent was twenty miles in length and twelve in breadth, and divided into three parts and nine lots; each child receiving one-third part, or three lots of the patrimonial estate. They were called, by way of distinguishing them from one another, 1st, the River or Water lots; 2d, the Long lots; 3d, the Back Short lots. The Water lots were bounded west by the Hudson, and were four miles square; the Long lots twelve miles in length from north to south, and four miles in breadth from east to west; the Short lots on the Connecticut line were the same size as the River lots—four miles square. Those lots acquired by Col. Robinson and Major Morris, by marriage with the two sisters of Capt. Frederick Philips, the father of Mrs. Mary Gouverneur, were confiscated by the legislature; but the reversionary interest was not affected thereby, which the late John Jacob Astor purchased of the heirs subsequently for $100,000; and for which, ten years afterwards, he received from the State of New-York $500,000, in State Stock at six per cent.

The following diagram shows the number, size, and form of these lots, and to which of the heirs they belonged. There is also a long, narrow strip of land, now the subject of litigation, which we have not designated in the diagram, between the north line of Putnam, as it now runs, and Rumbout's and Beekman's patents, claimed by the heirs at law of the original patentee.

HISTORY OF PUTNAM COUNTY.

Hudson River.

Col. Robinson's water lot, 4 miles square.	Capt. Frederick Philips's water lot, 4 miles sq.	Major Morris's water lot, 4 miles square.
Col. Robinson's long lot, 12 miles long, 4 miles wide.		
Major Morris's long lot, 12 miles long, 4 miles wide.		
Capt. Frederick Philips's long lot, 12 miles long, 4 miles wide.		
Major Morris's back lot, 4 miles square.	Capt. Frederick Philip's back lot, 4 miles sq	Col. Robinson's back lot, 4 miles square.

Left side: North line of Westchester Co.
Right side (top): Rumbout's Patent.
Right side (bottom): Beekman's Patent.

Connecticut line.

"Recorded for Mr. Adolph Philips—
"William the third by the grace of God King of England Scotland ffrance and Ireland Defender of the faith &c To all

to whom these Presents shall come Sendeth Greeting Whereas our Loving Subject Adolph Philips of our City of New York Merchant hath by his Peticon Presented unto our Trusty and welbeloved Benjamin Fletcher our Captain Generall and Governour in Chief of our Province of New Yorke and Territoryes Depending thereon in America &c Prayed our Grant and Confirmacon of a Certain Tract of Land in our Dutchess—Scituate Lying and being in the Highlands on the East side of Hudsons River beginning at a Certain Red Cedar Tree marked on the North side of the Hill Commonly called Anthonys Nose which is Likewise the North Bounds of Collonell Stevanus Cortlandts Land, or his Manour of Cortlandt and from thence Bounded by the said Hudsons River as the said River runs Northerly untill it comes to the Creek River, or Run of Water Commonly called and known by the Name of the Great fish kill to the Northward and above the said Highlands which is Likewise the Southward Bounds of another Tract of Land belonging (unto) the said Coll Stephanus Cortlandt and Company and so Easterly along the said Coll Cortlandts Line and the South Bounds of Coll Henry Beeckman untill it Comes twenty Miles or untill the Division or Pertition Line between our Colony of Connecticutt and our said Province and Easterly by the said Division Line being Bounded Northerly and Southerly by East and West Lines unto the said Division Line between our said Collony of Connecticutt and this our Province aforesaid the whole being Bounded Westward by the said Hudsons River Northward by the Land of Coll Cortlandt and Company and the Land of Coll Beeckman Eastward by the Pertition Line between our Collony of Connecticutt and this our Province and Southerly the Mannour of Courtlandt to the Land of the said Coll Cortlandt including therein a Certain Island at the North side of the said Highlands Called Pollepells Island which Reasonable Request we being willing to Grant Know ye that our Speciall Grace Certaine Knowledge and meere mocon We have Given Granted Ratifyed and Confirmed and by these Presents Do for us our Heirs and Successors Give Grant Ratify and Confirme unto the said Adolph Philips all the aforerecited Certaine Tract of Land and Island within the Limites and Bounds aforesaid together with all and Singular the Woods underwoods Trees Timber Hills Mountains Valleys

Rocks Quarreys Marshes Swamps Rivers Runs Rivoletts Waters Watercourses Pools Ponds Lakes fountains Streams Meadows fresh and salt Mines Minerals (Silver and Gold Mines Excepted) fishing fouling hunting and hawking and all other Royaltyes Rights Members Benefits Profites Advantages Commodityes Priviledges Hereditaments and Appurtenances whatsoever unto the aforerecited Certaine Tract of Land and Island within the limites and Bounds aforesaid belonging or in anyes Appertaining To have and to hold all the aforecited Certaine Tract of Land and Island within the Limites and Bounds aforesaid together with all and Singular the Woods Underwoods .Trees Timber Hills Mountains Valleys Rocks Quarryes Marshes Swamps Rivers Runns Rivoletts Waters Watercourses Pools Ponds Lakes fountains Streams Meadows fresh and salt Mines Minerals (Silver and Gold Mines Excepted) fishing fowling hunting and hawking and all other Royaltyes Rights Members benefits Profites Advantages Commodityes Priviledges Hereditaments and Appurtenances whatsoever unto the aforecited Certaine Tract of Land and Island within the Limites and Bounds aforesaid belonging or in any wayes Appertaining unto the said Adolph Philips his Heirs and Assignes to the sole and only Propper use benefite and behoofe of him the said Adolph Philips his Heirs and Assignes forever To be holden of us our Heirs and Successours in ffree and Common Soccage as of our Mannour of East Greenwick in our County of Kent within our Realme of England Yielding Rendering and Paying therefore Yearly and every Year unto us our Heirs and Successours forever at our City of New Yorke on the feast Day of the Annunciation of our blessed Virgin Mary the Yearly Rent of twenty Shillings Currant money of our said Province in Liew and Stead of all other Rents Services Dues Dutyes and Demands whatsoever for the said Tract of Land Island and Premises

"In Testimony whereof we have Caused the Great Seal of our said Province to be hereunto affixed Witnesse our Trusty and welbeloved Benjamin Fletcher our said Captaine Generall and Governour in Chiefe of our Province of New Yorke and Territoryes Depending thereon in America and Vice Admirall of the same our Lew : t and Commander in Chiefe of the Militia and of all the forces by Sea and Land within our Collony of Connecti-

and of all the forces and Places of Strength within the same cut in Councill at our ffort in New Yorke the seventeenth day of June in the ninth Year of our Reigne Annoy Dom 1697—
Ben ffletcher by his Excellencys Command
"DAVID JAMISON
"D. Secry

"I do hereby Certify the foregoing to be a true Copy of the Original Record Part of the word unto being interlined between the 26th and 27 lines of page 119 Compared therewith By Me
"LEWIS A. SCOTT Secretary."

ROADS AND TURNPIKES.

On examining the early records of Dutchess county, we find that the first road, in that part of Dutchess which is now Putnam county, was laid out, and the description thereof entered on the record, on the 28th day of April, 1744, by David Hustis and Francis Nelson, two of the commissioners appointed for that purpose. Afterwards several were laid out by Thomas Davenport, great-grandfather of William Davenport, Esq., of Nelsonville, and James Dickinson, who were also commissioners, some of which terminated in Westchester, Dutchess, and Connecticut. Many of these roads have been discontinued, or superseded by others more fitly located, some shortened, some extended, and some still remain but slightly altered. Generally their description is brief, imperfect, and obscure; and the different places mentioned in their description, with but few exceptions, are only known to the oldest inhabitants. Hustis, one of the commissioners, who signed the first entry on the record, made his mark; but whether from inability to write, or some infirmity of his hand, we have not been advised. But if inability to write was the reason, it is not to be wondered at, as the instance was more common then than now, and schools at that early day were, like angel's visits, "few and far between." We incline, however, to the belief, that it was owing to a palsied arm. In all extracts from the early records, either. county or town, we, to use a lawyer's phrase, "*stick to the record*" literally; adhering to the spelling and punctuation as it appears in the books, without addition, correction, or diminution of "one jot or tittle." The first entry on the record is as follows:

"Whereas by an Act of Generall assembly Passed in the Eleventh Year of his now Majesties Reign Entitled an act for the better clearing and further laying out public high Roads in Dutchess County: by Virtue of the same, We Francis Nelson and David Hustis being two of the Commissioners for laying out Roads in the South precinct in said county appointed have at the request of divers of the inhabitants laid out and ascertained the following public hig Ways or Road as follow viz, Beginning att Thomas Cercomes house from thence by marked Trees to - Epram Forgeson On Courtlandt's Manor,

"Another Road Beginning at the farm of Eli Nellson from thence by marked Trees to Nathan Lane's on the line of Courtlandt, thence down the line to the shrub plain—also one Road Beginning at the West Branch of Croton River at a Bridge— from thence by marked Trees down to Joseph Travers is—then running down the Dwivision Line through the still water to said Forgesons—One other road beginning at the deep brook or Roge Rill—from thence by marked Trees through pussapanun —thence to Daton's Hills; also One other Road Beginning at Hendrick Brewer's at pussapanun by marked Trees to Daton's Hills; One other Road Beginning at Sibet Cronkhyt at ye Indian Road Beginning at Joseph Jaycocks—from thence by marked

ROADS AND TURNPIKES. 85

Trees to the King's Road at Joseph Areles; one other Road Beginning below Pussattanun at Joseph Cronkhydt house by marked Trees to Datons Mill Performed by us said Commissioners the Twenty-eighth day of April in the Seventeenth year of his Majesties Reign Anno que Domini 1744.

" Dutchess ss   " DAVID × HUSTIS
                         mark
                  " FRANCIS NELSON
" A true Copy Examined
                  " by HENRY LIVINGSTON Clerk"

We remark—in order to show in which of the present towns of the county those persons resided, at whose houses the above-described roads commenced, and to determine the other points on their route, as far as we have been informed by old persons—that " Epram Forgeson" and " Thomas Cercome" lived in the now town of Carmel. " Eli Nellson" lived where Charles Smith now does, in Putnam Valley, and " Nathan Lane," where Robert Austin resides in the same town. The "*shrub plain*" is now called Hyatt's plain, and is about one mile below the Westchester line. " *A brige*," as mentioned in the entry, we have been informed, is now called Pine's Bridge. " Joseph Taveersis" lived a little west of the "*bridge.*" The " *still water*" was about six miles east of Peekskill. " Daton's Mills" are now known as Courtlandt's, and are about one mile east of Annsville. " Sibet Cronkhyt" lived between Annsville and St. Anthony's Nose mountain. " Joseph Jaycocks" lived near Annsville; and " Joseph Areles" lived where Reuben Turner does at present, a short distance north-west of Continental Village.

In 1745, three new commissioners were appointed to lay out highways, two of whom seem only to have

8

acted as such. They were " Adolph Phillips, Thomas Davenport, and James Dickinson ;" the first did not act. The descriptions or bounds of those first recorded by the clerk of the county are as follows:

"Whereas Adolph Phillips Esq. Capt Thomas Deavenport and James Dickinson Junr; are appointed Commissioners of ye high ways of ye South precinct of Dutchess County. We whose Names are here under subscribed being Two of ye Commissioners afforesaid have laid out ye high Ways, as is hereafter Mentioned—A high Way Beginning at the Devition line between Esqr. Philips pattain and Collonel Beekmans precinct near ye east part of ye precinct where ye: path is now Used so by marked Trees and Stakes threw ye precinct To Courtland Pattain; then a high Way from James Dickinson by Marked Trees to Courtland pattain; a high Way from James Dickinson by marked Trees to Rigfield new purchase, a high Way beginning at ye Devition of ye, Two countys near by Elihu Townsends at a White Oak Tree on ye East Side of ye high way from thence to a white Oak Tree—then to Elihu Townsend fence to his Corner as ye fence now stands then with ye, Middle line of ye Oblong untill it meets with Danbery high way by marked Trees from thence by marked Trees Over Joes Hill so called untill it meets with ye high way that comes from Wostershere so called; A high way Beginning at ye Bridge by John Dickinson so by marked Trees untill to Crane Mills from thence by marked Trees and Stakes to ye bridge by Jeremiah Calkins—A high way from Crane Mills by Marked To Rigfield New purchase—a high Way Beginning at Edward Grays so by Marked to ye Meeting house—from thence to ye West Branch of Croton by Marked Trees Meeting with ye highway that has been allready laid out near by Hamblins, a highway from Shaws by Marked Trees To Frost Mills from thence to Sprages, A High Way from ye Bridge by John Dickinson so by marked Trees to the Meeting house from thence by Marked Trees To Elijah Tomkins. A high Way Beginning Near by Taylors so by marked Trees Untill it meets with ye high way that comes over ye Great Swamp by William herns, A high Way by Marked Trees from William herns on ye North Side of ye Barr Swamp so called Untill it meets with Madam Britts Line A high Way by marked Trees from Madam Britts Line to the Horse pound so called from thence to Shaws and from ye horse pound To Croton River by Marked, A high Way beginning at Joseph Lees by Marked Trees to Wostershere high Way, A high Way from Sam'l Fields farme to ye high Way that leads to Danbery, A high Way from James Dickinson farme to ye high Way that leads to Courtland Line, A high Way from James Padocks To Connecttecut Line by Marked Trees A high Way Beginning at Capt. Balls possettion by Connectecut Line by marked Trees, A high Way beging at ye high Way near by Brundedges so by Marked Trees to William Bloomers—Give Under Our hands this tenth of May 1745

"Thomas Davenport
"James Dickinson, Jun'r.
"Dutchess County ss: A True Copy Examined
"By Henry Livingston, Clerk."
"August 23 : 1745."

"Crane Mills," of which mention is made in the above entry, were about half a mile from Sodom Corners, and on the north side of Joe's Hill, in the town of South-East. The "Horse Pound" was about three miles north of Carmel village, on the road leading from it to Stormville. The *Horse* Pond is on the same road, and the "pound" was immediately north of it.

"Shaw's" residence was in the town of Carmel, just north of the beautiful sheet of water that bears his name.

"*April ye* 20 *day* 1747: A highway Laid out Beginning at Abraham Smiths to by Marked Trees to the highway that Leads from Kirkun Mills to ye peakskills four Rods wide.

" A highway laid out Beginning at James Mairude So by Marked Trees to ye highway that leads from Kirkuns Mills to the peekkills four rods.

" A highway Laid out Beginning at a former Highway Near Ele

Nelson so by Marked trees to the former highway in peeckkills hollow four rods wide.

"A highway Laid out Beginning Near Mickell Shaw so by Marked Trees to the highway by Mathees Roes from thence by Marked Trees threw Mr hill farme to Kirkuns Mills four rods wide.

"A highway laid out beginning at Kirkuns mills so by Marked Trees to ye highway formerly Laid out that leads to the Peackkills four Rods wide.

"A highway laid out Beginning Near Benjamin Brundeges so by Marked trees to Josia Gregory four Rods wide.

"Dutchess ss: A True Copy examined August 19—1748. By Henry Livngston Clerk. } JAMES DICKINSON
THOMAS DAVENPORT"

At this period of the settlement of the county, there appears to have been but few roads, and they were scarcely worked. A journey, therefore, to Poughkeepsie, by the Commissioners, was something more of an undertaking than at the present day. Some half-a-dozen or more roads were laid out before the Commissioners, or one of them carried a description of them to the County Clerk to be recorded. This accounts for their entry under one date. The next entry is as follows:

"*March ye 20 day* 1746-7. A higway Laid out beginning at Kerkuns Mill so by Marked trees to peeks kill hollow from thence to Abraham Smith from thence to the highway that Leads Kirkun Mill to ye peeks Kill four rods wide.

"A highway Laid out beginning at Kirkuns Mill by marked trees to ye Highway to Eastward of Benjamin Brundages four Rods wide

"A Highway Laid out beginning at ye peeks Kill Road so by marked trees to Josia grigory four Rods wide.

"A Highway Laid out beginning at James Moreds to the peek Kill highway four Rods wide.

"Laid out by us Commissioners of ye Highways.

"Dutchess ss : A True Copy Examined By Henry Livingston Clerk April 8 1748-9.   } THOMAS DAVENPORT
JAMES DICKINSON

The descendants of those persons mentioned in the above-described roads, with those that follow, can better locate these early roads, from their knowledge of the places mentioned, than we. The next entry is as follows :

"*November ye 11 day* 1748. A Highway laid out from Capt: Wright Sawmill by marked trees to ye peac pond or to Westchester County Line four Rods wide

"A Highway Laid out from Curhelus fullers by marked trees Until it meets with the Road that Leads from ye Long bridge to Dan; l Grays four Rods wide

"A Highway Laid out from James Dickinson unto Court Lands maner by marked trees four rods wide Laid out by us Commissioners of the Highways

"A Highway laid out by marked Trees beginning at Croton River near James Dickinson from thence to ye high way by Lathams four rods wide

Dutchess: THOMAS DAVENPORT
                                   JAMES DICKINSON.

"A true copy examined by Henry Livingston Clerk Feb: 8: 1748-9—"

"Whereas ye Inhabitants on ye South Precinct in Dutchess County in the province of New-York Did Request Severall Highways To be Laid out wee ye Said Commissioners have laid out Several Highways as follows fiirst Begining near James Dickenson, from Thence by Marked Trees To Courtland Maner by Nathan Balys four Rods wide

"Then one more High Way Beginning by whare Doctor Calkins Used to live from Thence by Marked Trees To ye Oblong.

Thence Between Nathaniel Stevenson and Philips pattain To Beekmans precinct four Rods Wide. One more highway Beginning near More Houseis Mill by marked Trees to ye Old High Way and ye: Old High way by Greenes House Stopt up four Rods wide.

"One more High Way Beginning near by Joseph Cranes from thence by marked Trees into ye High Way by Saml: Jones four Rods.

"One more Highway Beginning at the South End of Nathaniel Stevensons Land from Thence East-ward in Between Stevensons Land Joshua Burns Land four Rods Wide to ye Middle of the Oblong

"Laid out by us
    THOMAS DAVENPORT
    JAMES DICKINSON
   Commissioners of the Highways."

"Whareas ye Inhabitants on Phillips pattain have Requested A Highway to by laid Out from Timothy Shaws to ye Fish Kills Through ye Mountains or over ye Mountains Which We have Done Beginning at Timothy Shaws Aforesaid four Rods Wide by Marked Trees to the Fish Kills as aforesaid by us Commissioners of the HighWays for ye south precinct of Dutchess County.
    THOMAS DAVENPORT
    JAMES DICKINSON
   Commissioners of the High Ways

"Dutchess ss: The above are True Copys Examined by Henry Livingston Clerk
June 5: 1752—"

"A High Way Laid out Begining at Jonathan Lanes House from Thence by Marked Trees to Elezer Umans Mill four Rods Wide, A High Way Begining at Timothy Shaws from Thence Over ye Mountains To the fish Kills by Marked Trees four Rods Wide Laid out by us Commissioners of ye highwas of ye South precinct in Dutchess County June ye, 7th: day 1751—.
  "THOMAS DAVENPORT
  "JAMES DICKINSON

"Dutchess: A True Copy Entred January 8: 1775
   "By HENRY LIVINGSTON Clerk"—

"November ye : 10 day 1752 South precinct of Dutchess County A high Way Laid out from Amos Dickinson to Jeremiah Jones by Marked Trees 4 Rods Wide One more Beginning at ye Horse pound from thence to Amos fullers 4 Rods Wide by Marked Trees, One More Beginning at John Dickensons Mill from thence to ye : high Way that Leads to the Meeting house 4 Rods Wide

"Thomas Davenport
"James Dickinson
"Comistioners of the high Ways—

"Dutchess : A True Copy Recorded January 8 1755 By
"Henry Livingston Clerk"

"October ye 11 : day 1754 South precinct of Dutchess County, A high Way Laid out Beginning at ye Bridg Near Edward Halls Mill on ye : Oblong from thence by John Ryder door to a Stake in said Ryders Meadow from thence between James Anderson Land and said Ryders Land as far as is Convenant for a high Way to be made from thence as near to Rattle Snake hill as is Convenant for a high Way to be made from thence to the highway that Leads across Joes Hill so called Two Rods Wide Throughout One more beginning at the high Way that Leads to Roberts paddricks on the Top of the hill in John Jones Possestion from thence by Marked Trees to Jacob Finch Bridg from thence by James Quimby And from thence to Thomas Frost 4 Rods Wide One more Beginning on ye West Side of Quimbe farm at ye highway from thence between John Frost And James Quimbe farms And from thence to Thomas Townsend And from thence to the Bridg by Jeremiah Baleys 4 Rods Wide One more Beginning near Nehemiah Woods at ye : high Way from thence to Nathaniel Byingtons Bridg four Rods Wide

"One more Beginning at Thomas Higins from Thence a Crost ye: hills to Daley Brook so called 4 Rods Wide by marked Trees One more Begining at Anthony Batterson House from thence along ye: Collony Line to ye: highway that leads to Danbury. 2 Rods Wide

"One more Beginning at ye foot of the hill near ye: peach pond from thence by Marked Trees and Bushes and Stakes and

Stones to West Chester Line four Rods wide Throughout One more Begining at ye: foot of a hill in Mathew Burgis Land by Marked to ye: Top of hill in to ye: Old highway Again 4 Rods Wide Laid out by us

"Thomas Davenport
"James Dickinson
"Commistioneers of the High Ways

"Dutchess: A True Copy Recorded January: 8: 1755 By
"Henry Livingston Clerk"

REVOLUTIONARY LETTERS, &c.

(Letter from Col. Ludington, Elijah Townsend, and others.)

"Dutchess County, 3d December, 1776.

"Gentn.—Nothing but the strongest necessity could induce us to trouble you with an application of so extraordinary a nature; but if we are esteemed worthy your confidence as friends to our struggling country, our sincerity will apologize for what in common cases might appear indecent. Our invaded State has not only been an object of the special designs of our common enemy, but obnoxious to the wicked, mercenary intrigues of a number of engrossing jockies, who have drained this part of the State of the article of bread to that degree, that we have reason to fear there is not enough left for the support of the inhabitants. We have for some months past heard of one Helmes who has been purchasing wheat and flour in these parts for several months, with which the well affected are universally dissuited. This man with us is of doubtful character, his conversations are of the disaffected sort entirely. He has now moving from Fishkill toward Newark we think not less than one hundred barrels of flour, for which he says he has your permit, the which we have not seen. However, we have, at the universal call of the people, concluded to stop the flour and Helmes himself, until this express may return. We ourselves think from the conduct of this man that his designs are bad.

"We have the honour to be, your humble servts.

"HENRY LUDINGTON,
"JOSEPH CRANE, Junr.
"JONATHAN PADDOCK,
"ELIJAH TOWNSEND."

"To the Honourable the Council of Safety for the State of New York."

"Fredericksburgh Committee.
March 15th, 1776.

"Whereas Isaac Bates has been represented to this committee as being unfriendly to our country, we have had him under examination and find him guilty of said charge. We, therefore, refer him to the Honourable county committee for further examination."

"Fredericksburgh, March 15th, 1775.

"Isaac Bates, upon being taken up as a deserter, by an advertisement from Elijah Oakley, Lieutenant under Captain Comfort Ludington, of Colo. Jacobus Swartwout's regiment of minute men, pleads and says that said Lieut. Oakley did release him, in support of which plea he produced the evidences, whose depositions are as follows:

"I, Abraham Birdsil, of lawful age, being sworn before the chairman of the committee, do testify and say that on the 5th of this instant March, being at the house of Cornelius Fuller, I heard Elijah Oakley say he would give any man two shillings that would set his name to such a paper. Whereupon Isaac Bates said he would set his name to it; and the said Oakley said he would give him four shillings if he would; and finally said as he could not make change he would give him a six shilling bill, lawful money. And as Bates took the pen Oakley says if you do write your name there you shall go, and Bates said I mean to go, and wrote on the bottom of the paper as I supposed his name, but I understand by others (for I cannot read writing) that he wrote Elijah Oakley may kiss my —— Isaac Bates; at which Oakley was mad and swore he should go. Whereupon Bates says why you are not mad are you, I was only in a joke. Joke or no joke said Oakley, you shall go But afterwards I saw Bates give Oakley the bill again, and saw Oakley tear off a piece of paper which I suppose was what Bates had written, and I understood by Oakley that he had discharged him. Whereupon I said to Bates, since Oakley is so fair with you, you ought to treat him, and he immediately called grog and did treat him."

"I, John Chase, of lawful age, being sworn before the chairman of the committee, do testify to the whole of the foregoing deposition; and further that when Oakley took the bill he said he would see if it was the same bill which he gave Bates, and went to the light and said it was the same bill which I gave you. Now (said I to Mr. Oakley) you and Isaac are clear, are you not? Yes, said Mr. Oakley we are clear, it was only a joke."

"We do suspect the above mentioned Elijah Oakley as being unfriendly to the country, from his conduct in enlisting Isaac Bates who was known to be a professed tory, and taking him out of our hands when we were about to deal with him, and then discharged him, but at the same time positively affirmed to us that he would make him go, and finally did advertise him, when he never kept out of his way.

"By order of the Committee of Fredericksburgh,
"DAVID SMITH, Chairman."

"March 16th, 1776.

"4 ho. 6 m. March 28th, 1776.

"The Committee met pursuant to adjournment.
"Present—Mr. Wm. Paulding, Chairman.
"Mr. Cuyper—Orange.
"Mr. Moore—Tryon.
"Mr. Everson, Colo. Morris Graham—For Dutchess. Mr. Lefferts—Kings. Wm. Williams—Cumberland.
"Mr. Tredwell—For Suffolk.
"Mr. Paulding—For Westchester.
"Mr. Ad. Bancker—Richmond.

"A return for a great number of Commissions from Fredericksburgh, in Dutchess County, for the militia officers in that district, was read and filed, and is in the words following, to wit:

Fredericksburgh in Dutchess County, March 15th, 1776.

"Pursuant to a resolve of the Provincial Congress of New York, passed the 9th of August, 1775, the Committee proceeded to call together the several companies of militia in this precinct, for a choice of officers, as follows:

"Beat No. 1. Friday March 8th, the company did meet, and under the inspection of Joshua Myrick, Daniel Mertine, and David Myrick, three of the committee, did choose Ebenezer Robinson, Capt.; Nathaniel Scribner, 1st Lieut.; Hezekiah Mead, Junr. 2d lieut.; Obadiah Chase, Ensign.

"Beat No. 2. Monday March 11th, the company met, and under the inspection of David Waterbury and Moses Richards, two of the Committee, did elect David Waterbury, Capt.; Isaac Townsend, 1st lieut.; Jonathan Webb, 2d lieut.; Timothy Dela van, Ensign.

"Beat No. 3.—September, 20th, 1775, the company met, and under the inspection of Jonathan Paddock, Simeon Tryon, David Crosby, three of the Committee, made choice of Jonathan Paddock, Capt.; Jeremiah Burges, 2d lieut.; Joseph Dykeman, Ensign.—N. B. Simeon Tryon is since appointed a lieutenant in the Continental Army.

"Beat No. 4.—Tuesday, March 12th, the company of -- -- met, and under the inspection of Solomon Hopkins, David Myrick, and David Smith, did elect John Crane, Capt.; Elijah Townsend, 1st lieut.; David Smith, 2d lieut.; and John Berry, Ensign.

"Beat No. 5.—Wednesday, March 13th, the company met and under the inspection of Solomon Hopkins and Joshua Myrick, two of the Committee, did elect Wiiliam Colwell, Capt.; Joel Mead, 1st lieut.; Stephen Ludinton, 2d lieut.; and David Porter, Ensign.

"Beat No. 6.—Thursday, March 14th, the company met, and under the inspection of Isaac Chapman and Joshua Crosby, two of the Committee, did choose David Hecock, Capt.; William Calkin, 1st lieut.; and Moses Sage, ensign.

"The above gentlemen are all persons of respectable characters, have been friendly to liberty, and have signed the general association recommended by the Congress.

"By order of the Committee,
"DAVID SMITH, Chairman, *pro tempore*.

"A true copy, Test.
"JOSHUA MYRICK, Clerk.

"N. B. Increas Bennet afterwards refused to serve as lieutenant."

"Die Sabbati, 9 ho. A. M.
"July 20th, 1776.

"The convention met pursuant to adjournment. Opened with prayer.

"A letter from Col. Henry Ludenton, of Dutchess County, dated the 19th instant, was read and filed. He thereby informs that there are many vacancies of Captains and subalterns in his regiment, besides, that the offices of 1st and 2d major are also vacant. He recommends, with the advice of the precinct committee, Mr. Robinson and Mr. Gee for majors, and requests 20 blank commissions by the bearer, who says part of the regiment is to march to-morrow, and that they have delayed for commissions.

"On reading the said letter from Col. Ludenton, of Dutchess County, and considering the state of his regiment at this critical time, Resolved, That commissions be issued to the two gentlemen therein named, as majors of that regiment, in the order they are named in the said letter, and that twenty other commissions be signed by the President, and countersigned by one of the Secretaries, and transmitted to Colo. Ludenton, to be filled up for the Captains and subalterns of his regiment, when necessary, by the precinct committee and himself; and that the said precinct committee and Colonel Ludenton do return to this Convention an exact list of the names, rank and dates of the officers, commissions which they shall fill up and deliver. And Resolved, That the sending blank commissions to a precinct committee shall not, from this instance, be drawn into precedent.

"A draft of a letter to Colo. Ludenton, was read and approved, and is in the following, to wit:

"SIR—Agreeable to your request in your letter of yesterday,

we now send you 20 blank commissions to be filled up by you, in conjunction with the committee of the precinct, for the captains and subalterns wanted in your regiment; and have likewise enclosed two commissions, appointing Mr. Robinson and Mr. Gee majors. As the Congress were not informed of Mr. Gee's christian name, you, together with the Committee of the precinct are requested to insert it.

"By order.

"To Colo. Henry Ludenton."

"Die Veneris, 10 ho. A. M. Novr. 22d, 1776.

"The committee of Safety met pursuant to adjournment.

"Mr. William Duer informed the Committee that large quantities of hay and corn were purchased by the Quarter Master-General for the use of the Continental army in the eastern parts of this county and the western parts of Connecticut, and that it would be hardly practicable to convey the same to the army unless the roads leading from the Oblong and Frederickburg towards Reze's Bridge and North Castle were better repaired; he therefore, in behalf of Gen. Mifflin, Quarter-Master-General of the Continental army, prayed that this House would devise ways and means of facilitating the above mentioned communication, not doubting but so necessary an expenditure would be cheerfully reimbursed from the Continental Treasury.

"On taking the application of Mr. Duer into consideration, Resolved, That it will be necessary to repair the following roads in order to facilitate the cartage of forage to the Continental army; from the house of John Miller towards the house of Colonel Henry Luddington, thence to Samuel Washburn's, being eight miles; the road which runs east from Colonel Henry Luddington's to the store of Malcolm Morrison, and thence south to the mills of Samuel Washburn, being twelve miles.

"Resolved, That Colonel Luddington detach from his regiment one hundred men for the purpose of repairing that part of the road which is first mentioned, being in distance 8 miles.

"Resolved, That Capt. H. Meade be appointed superintendent for repairing the above roads.

"Resolved, That Colonel Field detach one hundred men from his regiment of militia for the purpose of repairing that part of the road which is last mentioned in the first resolution, being in distance 12 miles.

"Resolved, That Capt. David Heacock be appointed superintendent for repairing the said road."

"A letter from the Committee of Dutchess County, was read as follows, viz :

"Dutchess County, May 6th 1776.

" Sir—It having been represented to the general committee of this county, that the southern regiment of militia was too large and extensive, containing twelve companies, and covering a space of country upwards of thirty miles in length, we have therefore, not only because in other respects it was expedient, but also in compliance with the resolution of Congress prohibiting a regiment to consist of more than ten companies, divided it, and instead of one have formed the militia in that quarter into two regiments. Enclosed you have the description of the regiments, together with a list of persons nominated for field officers. As this part of our militia will remain unregimented till the officers receive their commissions, we must request that the commissions be made out as soon as possible, and sent to the Committee in Rumbout's precinct, with directions to forward them to the officers immediately.

"I remain, (by order of the committee,) your very humble servant. EGBERT BENSON, Chairman."

"The description of the two regiments, enclosed in the letter from the committee of Dutchess, was read as follows:

" One regiment, to consist of all the militia in Pauling's precinct, (except the northern company,) all the militia in Southeast precinct, and the militia on the northern and middle *short lots*, in Fredericksburgh precinct, in the county of Dutches. John Field, Colonel; Andrew Morehouse, lieut. Col.; Jonathan Paddock, 1st Major; Isaac Tallman, 2d major; Isaac Crane, adjutant; Reuben Crosby, quarter-master.

"The other regiment to consist of all the militia in Fredericksburgh precinct, (except the northern and middle short lots,) and all the militia in Phelps precinct, (this should have been written 'Philips precinct,') in the county of Dutchess. Moses Dusenberry, Colonel; Henry Luddington, lieut. Col.; Reuben Ferris, 1st major; Joshua Nelson, 2d major; Joshua Myrick, adjutant; Solomon Hopkins, quarter-master."

THE OBLONG.

This is a tract of land one and three-fourths of a mile in width, commencing in the town of Rye, in Westchester county, and running north through it, Putnam, and Dutchess; the west side of which was, until 1731, the boundary line between New York and Connecticut. The east side of it now forms the division line of the above-named States. It contains 60,000 acres, and just before the death of Col. John Montgomery, in 1731, who succeeded William Burnett, in 1728, as Governor of the Province of New York, it was ceded to this State in consideration of another tract near Long Island Sound, surrendered to Connecticut. This State drew a line through its centre, divided it into 500 acre-lots, and sold it to emigrants, who received a guaranty of title from the State. It was this security of title which caused these lots to be eagerly sought after by emigrants from Cape Cod.

REVOLUTIONARY PLEDGE.

On the 18th day of April, 1775, a detachment of British troops under Colonel Smith, was sent from Boston by General Gage, to destroy some American stores collected at Concord, then a small village, six miles north-west of Lexington, in Massachusetts. Upon Lexington Common seventy men were drawn up, on whom Major Pitcairn ordered the detachment to fire. The order was promptly obeyed, and seven men were killed and three wounded. There the blood of patriots was first shed, that was to nourish the infant tree of Liberty, during a seven years' struggle, while the ruthless elements of tyranny were warring for its destruction. On the 29th day of the same month and year, and eleven days after the bloody tragedy at Lexington, the inhabitants of the city of New York called a meeting of all who were opposed to the oppressive acts of the English Parliament, formed a general association, adopted a *Pledge*, and transmitted a copy to every county in the State for signatures.

The storm had burst, and every day was adding fearful intensity to its force.

The proud Lion of England had lapped the heart's blood of the descendants of the Plymouth-Rock Pilgrims; and their brethren of the other colonies saw that, ere long, with a few more bounds, he would leap among them. Between submission and resistance they were called to choose; the former they had yielded to until it had ceased to become a virtue, and the latter was the only alternative left to men who were determined to wear the yoke no longer. The British Parliament and King had as zealous partisans and friends among us, as they had at home. It became necessary, in some way, to ascertain who were the friends of our *own*, and the mother-country. The Pledge was suggested; and, acting on a test of divine origin, they who refused to sign it were set down as opposed to their country and the maintenance of her rights. In order to secure unanimity of purpose and harmony of action—to ascertain who could be relied on in the different counties, and draw out their political sentiments on the issue joined between the *unnatural Mother* and her *rebelling Daughters*—to commit the people to one side or the other of the question, and by united action among the friends of the cause, to prepare for the approaching conflict, was the object of the Pledge. If there ever was a "time that tried mens' souls," it was when they grasped the "gray-goose quill" to sign their death-warrant if they failed, or their liberty if successful. At that period, even the most violent patriot must have looked upon the undertaking as desperate and almost hopeless, with but one chance out of ten in his favor. But they were men of a by-gone and an iron age, upon whom the world may not look again. They had made up their minds to die rather than submit; and when men of such indomitable energy of mind once deliberately resolve, their destiny is fixed. New York, it will be seen, moved early to ascertain, after hostilities had commenced, the sentiments of her citizens on the issue of a nation's freedom. The Pledge was as follows;

"Persuaded that the salvation of the rights and liberties of America depend, under God, on the firm union of its inhabitants in a rigorous prosecution of the measures necessary for its safety; and convinced of the necessity of preventing anarchy and confusion, which attend the dissolution of the powers of government, we, the freemen, freeholders, inhabitants of ―――, being greatly alarmed at the avowed design of the Ministry to raise a revenue in America, and shocked by the bloody scene now acting in Massachusetts Bay, do, in the most solemn manner, resolve never to become slaves; and do associate, under all the ties of religion, honor, and love to our country, to adopt and endeavor to carry into execution whatever measures may be recommended by the Continental Congress, or resolved upon by our Provincial Convention for the purpose of preserving our Constitution, and opposing the execution of the several arbitrary Acts of the British Parliament, until a reconciliation between Great Britain and America on constitutional principles (which we most ardently desire) can be obtained; and that we will in all things follow the advice of our General Committee respecting the purposes aforesaid, the preservation of peace and good order, and the safety of individuals and property."

"DUTCHESS COUNTY."

"Agreeable to adjournment to this day, being the 15th of August, 1775, we met at the house of Jacob Griffin, in order to make a return of the persons who signed the Association and those who refused, viz : Those who signed—

Theods. Van Wyck,	Dirck G. Brinckerhoff,
John Brinckerhoff,	Daniel Ter Boss,
Zachs. Van Vorhees,	Richard Van Wyck,
Garret Storm,	William Van Wyck,
Cornelius Sebring,	Joseph Horton,

REVOLUTIONARY PLEDGE. 103

Johannes Wiltse,
Gores Storm,
T. Van Wyck, Jr.,
Harvey M. Morris,
Henry Godwin,
Thomas Storm,
John Adriance,
Henry Schenck,
Jacob Swartwout,
Corns. Van Wyck, Jr.,
Isaac Sebring,
Abm. Brinckerhoff,
Roelef Schenck,
Abraham Schenck,
L. E. Van Bunschoten,
Isaac Ter Boss,
Jacob Griffin,
James Snediker,
Aaron Brown,
John A. Brinckerhoff,
John Wickoff,
James Denton,
William Clauker,
George Brinckerhoff,
Adrian Brinckerhoff,
Abraham Ter Boss,
John H. Sleght,
Jacobus De Graef,
John Meyer,
John G. Brinckerhoff,
John Langdon,
George Adriance,
George Elsworth,
Hendrick Boerum,
Daniel Schenck,
Jonathan Langdan,
William Tisdale,
Joseph Griffin,
Daniel Johnson,

John Meynema,
Abm. Van Voorhis,
Hendk. Hardenburg,
Moses Bedell,
Peter Ter Bush,
John Jewell, Jr.,
Alexander Turner,
James Auning,
William Ward,
Jacob Du Bois, Jr.,
Gabriel Hughson,
David Barker,
Henry Van Tessel,
Claistian Du Bois, Jr.,
Ahas. Elsworth,
Jacob Brinckerhoff,
William Holms,
Thomas Ostrander,
Godfrey Heyn,
N. E. Gabriel,
Abraham Morrell,
Geo. J. Brinckerhoff,
Christopher Rawn,
James Weekes,
Isaac Van Wyck,
Cornelius Smith,
Hugh Conner,
Andw. J. Lawrence,
Nathl. Fairchild,
Samuel Gosline,
John Berray,
James Cooper,
John Cooper,
James Barnes,
John Ter Bush,
Cornelius Adriance,
Abm. De Foreest,
Thomas Simonton,
Joseph Mc Cord,

John Cooper,
Richard King,
Jacob Van Voorhis, Jr.,
Jonathan Haight,
Israel Kniffin,
Daniel Kniffin,
Jonathan Kniffin,
Walter Heyer,
Adrian Bogert,
Moses Akerly,
Luke Ter Boss,
James Miller,
Cornelius Osborne,
Nicholas Brower,
Matthias Clark,
Nicholas Brower, Jr.,
John Wright,
Charles Brewer,
John Ackerman,
John Walters,
James Rathbun.
Seth Chase,
Adolphus Brower,
David Brower,
Cornelius Brower,
Jacob Brower,
Deriah Hogland,
William Haskin,
Peter Horton,
Jesse Bedell,
Martin Schenck,
Peter Monfoort,
Matthias Horton,
Johans. De Witt, Jr.,
Mat. Van Bunschoten,
Abm. Van Wyck,
Steph. Brinckerhoff,
Geo. Brinckerhoff,
John Scouten,
Joseph Balding,
J. Scouten, son of Jerry,
Jacobus Emans,
James Brown,
Moses Barber,
Abm. L. Losee,
Samuel Swartwout,
John Swartwout,
William Scouten, Jr.,
Daniel Rayner,
Robert Brett,
John Smith,
Jacob Balding,
Caleb Cornell,
Isaac Storm,
Henry Rosekraus,
Benjamin Rosekraus,
Stephen Osborne,
Simon S. Scouten,
Daniel G. Wright, Jr.,
Joseph Wiltse,
Geo. Van Werkeren,
Platt Rogers,
Theods. Adriance,
Micah Rogers,
John Lawrence,
Jeremiah Bedell,
Joseph Fowler,
Jacob Swartwout,
Gideon Way,
Merinus V. Vlaikren,
Henry Ostrander,
John Leyster,
Timothy Saikryder,
Zachariah Boss,
John Bush, Jr.,
Josiah Hallstead,
Peter Noorstrant,
Jeremiah Martin, Jr.

REVOLUTIONARY PLEDGE.

Peter Snyder,
John Gray, Jr.,
Gershom Martine,
Amos Nettleton,
John Bennitt,
Elihu Emmitt,
Ab. H. Van Amburgh,
Jesse Baker,
James Thurston,
Joseph Parker,
Stephen Thaiker,
Abraham Gray,
John Baker,
Jeremiah Ranny,
David Mowry,
Joseph Lee,
Simon Bise,
William Lane,
Ezra Mead,
James Innes,
Isaac Smith,
Peter Hulst,
David Bennett,
David Horton,
William Wright,
Daniel Canfield,
Sabure Main,
Johans. Brinckerhoff,
Andw. Van Hyning,
Abm. Van Amburgh,
Moses Saikryder,
James Rosekraus,
Stephen Doxey,
Dirck Hegerman,
Jonathan Talmagee,
Solomon Saikryder,
Joshua Hicks,
Martin Smith,
Robert Rogers,
Thomas Wright,
William Baker,
Daniel Wright,
John Watts,
Johans. De Witt,
Albert Carley,
Henry Van Voorhis,
Martin Wiltse,
H. Rosekraus, Jr.,
James Kilburne,
Dirck Brinckerhoff,
Zebulon Southard,
Evert W. Swart,
John Bloodgood,
Walter Moody, Jr.,
John Johnson,
Simon Ter Bush,
Thorn Pudney,
Francis Pudney,
Abraham Ceasa,
Stephen Peudy,
Henry Carpenter,
John Ter Bush,
Abraham Schultz,
Cornelius Sebring,
John Pudney,
Cornelius Ter Bush,
David Lyons,
Edward McKeeby,
Theods. Brett,
John McBride,
Obadiah W. Cooper,
Timothy Mount,
Jonas Southard,
James Reynolds,
George Bump,
Tunis Du Bois,
James Green,
Obadiah J. Cooper,

Peter Klump,
Abm. Van Tyne,
Jacob Van Voorhis, Jr.,
Myndert Cooper,
John Runnels,
Thomas Bump,
Christopher Schults,
Silvinus Pine,
Isaa H. Ter Boss,
William Somerdike,
Philip Pine,
Nathan Bailey,
John Pullick,
Austin Fowler,
David Pellet,
John Southard,
Duncan Graham,
Elesa Du Bois,
James Duncan,
Caleb Briggs,
James Osburn,
Isaac Hegeman,
Jacobus Degroff,
E. E. Van Bunschoten,
John De Groot,
Jno. Van Bunschoten,
Robert Jodd,
Bernd. J. Van Kleek,
Jabobus De Gruff, Jr.,
Jacobus Sleght,
Moses Vanelin,
Adam Dates,
William Stanton,
William Teatsort,
Isaac Snider,
Thomas Lewis,
Jacob Cole,,
Abraham Sleght,
Michal Hoffman,

Teunis Wilsen,
Isaac Cole,
Peter Stienbergh,
Gideon Ver Velon,
Moses De Groff,
Henry Buys,
Peter Van Kleek, Jr.,
Jeremiah Mead,
Henry Pelts,
Jacob Backer,
Jacob Coapman,
Barent Dutcher,
Boltes B. Van Kleek,
John Leroy, Jr.,
Henry Bell,
Jurrie Hoffman,
Jacob Niffer,
P. Van Dervoort, Jr.,
Simon Leroy, Jr.,
John Leroy,
Jacob Lane,
Thomas Yeumans,
Constine Gulnack,
Johans. Hooghteling
Clement Cornwell,
Peter Deets,
Francis Leroy,
Abm. Westervelt,
Jost. Westervelt,
James Howard,
Cornelius Griffin,
William Griffin,
James Vandewater,
Dalf Swartwout,
Garret Beneway,
Jeremiah Var Velen,
Thomas Pinkney,
Henry Marten,
Barthol. Hogeboom,

REVOLUTIONARY PLEDGE. 107

Charreik Van Keuren,
David Dutcher,
Deminicus Monfoort,
James Rymden,
Andrew Ostram,
John Ostram,
Frederick Rosekraus,
Peter Van Dewater,
Bareut B. Van Kleek,
Sevaris Van Kleek,
Francis Van Dewater,
John Van Valin,
Peter Polmetier,
Lawrence Conklin,
Herman Rynden,
John Rosekraus,
Thomas Johnson,
Francis Way, Jr.,
Joshua Smith,
Aaron Brown, Jr.,
Abraham Ladu,
Cornelius Swartwout,
Gilbert Lane,
William Swartwout,
James Swartwout,
Samuel Roberts,
Ebenezer Clark,
William Lane,
Joseph Totten,
Andrew Hill,
Johannes Sharrie,
Jeremiah Jones,
Lawrance Haff,
Peter Outwater,
Daniel Outwater,
T. Van Benschoten,
Samson Smith,
Albert Terhum,
Abm. Duryee, Jr.,

John Tirhum,
James Culver,
Dennis Culver,
James Culver, Jr.,
Peter Van Benschoten,
Jacob Van Benschoten,
Henry T. Wiltsey,
John Tappen,
James Davison,
Henry Burhause,
William Hogelandt,
Abijah Pattersan,
Daniel Terhum,
Abraham A. Lent,
Tunis Skeet,
Cornelius Verwie,
Hugh Laughlin,
Francis Hegeman,
John Culvert,
Abraham Cronckheit,
John Jewell,
Isaac Jewell, Jr.,
Cornelius Wiltse,
Hemming Higby,
Peter Lent,
Isaac Adriance,
Johannes Boss,
Richard Griffin,
Steph. Van Voorhis,
Jacob Buys, Jr.,
John L. Losee,
Jacob Horton,
Corns. Ostrander,
Richard Comfort,
Abraham Shear,
William Barnes,
Frederick Scutt,
Jerome Van Voorhis,
Kam Adriance,

Kam J. Adriance,
John Devoe,
Jac's C. Swartwoudt,
Peter Robinson,
Moses Shaw,
Jacobus Van Dewater,
Zach. Van Voorhis, Jr.,
William Brock,
Jacob King,
John Hutchins,
John Darlon,
James Wildee,
William Wildee,
Richard Avery,
John Rosekians,
Isaac Hutchins,
John Yurkse,
Abm. Van Wackere
Jacob Hutchins,
Thomas Way,
Abm. De Witt,
John Philips,
Elbert Mumfort,
Danl. Van Voorhis,
George Jewell,
John Noorstrant,
Peter Schoonhove,
Joshua Griffin,
Isaiah Wilde,
Isaac Southard,
William Winslow,
John Griffin,
John Vandervoort,
Daniel Shaw,
Peter Fitz Simmons,
Nathan Burnes,
John Vermillie,
Richard Osborn,
Peter Johnson, Jr.,

Richard Jewell,
Jacob Dubois,
Jacob Van Dervoort,
Peter Meyer,
John Coffin,
Coenradd Appleye,
Joshua Bishop,
William Van Tyne,
Sylvester Bloom,
John Van Sulen,
John Kipp,
William Brooks,
Jacob Van Tassell,
Stephen Bates,
Daniel David,
Isaac Griffin,
Peter Montross,
Isaac Holmes,
Aaron Shute,
Richard Jackson,
Dirick Hardenburgh,
Peter J. Monfoort,
Timothy Talman,
Peter Depung,
William Cushman,
Garret Handenburgh,
Tobias Mabie,
John Bogardus,
Samuel Somes,
Nathan Somes,
Jonathan Terry,
Ralf Phillips,
Isaac Jewell,
George Bloom,
Benjamin Roe,
Henry Hains,
Lawrance Lawrance,
Jonas Cauniff,
Edward Churchill,

Samuel Tedd,
William Roe,
James Miller,
John Phillips,
Daniel Auning,
Daniel Ward,
William Barker,
John Parks,
Peter Bogardus, Jr.,
John Davis,
William Earls,
Peter Bogart,

Francis Way,
William Fowler,
Corns. Brinckerhoff,
Dennis Mc Shebeey,
Isaac Veal,
Robert Mc Cutchin,
Robert Nichkilson,
Elias Concklin,
Jesse Purdy,
Joseph Ogden,
Andrew Renvells,
William Ardem.

A list of persons in Dutchess County who refused to sign the Association:

Of Captain Heganan's Company:

John Tarpanning,
John Jast Snider,
John Crandle,
James Medagh,
Michas Cock,
Barent A. Van Kleek,
William Baker,

Urean Terwilger,
Jores Middagh,
Daniel Cole,
Albert Monfoort,
Henry Cailen,
Peter Burhans,
William Rogers.

Captain Stephen Brinkerhoof's Company:

John Holsted, Lieut.,
Jacob Wright,
Joseph Morss,
Benjamin Snyder,
Oliver Peck,
William Cure,
Joseph Ferinton,
Joseph Merritt,
Johannes Devoe,
Joseph Smith,
Joseph Robison,
Ebenezer Pellit,

Peter Boss,
Richard Yeats,
Jessey Baker, Jr.,
Christopher Winter,
Benjamin Ellis,
Joseph Halsted,
Thomas Martin,
John Miller, Lieut.,
  of Captain Lodinton,
Ezekel Main,
Levi Winter,
Joseph Winter,

Tunis Kranchite,
William Goodfellow,
Isaac Wright,
Benjamin Doty,
Jonathan Lee,
Benjamin Ogden,
David Roe,
Joshua Odle,
Semeon Losee,

Philip Roens,
Thomas Baker,
Johannes Storm,
Stephen Stolker,
Philip Morse,
Daniel Haasbroock,
Thomas Carman,
Zebulon Gray,
Silas Brown.

Captain Southard's Company:

Gerrèt Nostrand,
Johannes Voorhees,
Abraham Philps,
Henry Philps,
Peter Philps,
Jocobus Philps,
Elias Van Voorhees,
Richard Southard,
Thomas Southard,
Gilbert Southard,
Richard Southard, Jr.,
Daniel Southard,
Thomas Poyer,
Robert Bogardus,
Increase Mills,
Robert Mills,
Henry Mills,

Samuel Mills,
Jesse Purdy,
Joseph Green,
Francis R. Britt,
Jeremiah Cooper,
Jonas Halsted,
Jacob Rider,
John Covert,
Leniah Adams,
Philip Shoaf,
Thomas Gibson,
Peter Brogardus,
Isaac Vealey,
Thomas Sprage,
Jeremiah Green,
Benjamin Munger,
Thomas Miller.

Captain John Bedle's Company:

John Bedle, Captain,
John Schutt, Lieut.,
F. Hasbrook, Lieut.,
George Van Nostrand, Ensign,
James F. Way,
Endrew J. Schouten,
Benjamin Gerox,
John Linabeck,
Jacobus Jno. Schutt,

Enoch Purdy,
Joseph Burroughs,
John S. Langdon,
Joseph Wood,
Joseph Carey,
Isaac Wood,
Oliver Larduex,
Thomas Craft,
Peter Dubois,

REVOLUTIONARY PLEDGE. 111

Nathaniel Laduex,
Johannes Swartwout,
Joseph Winn,
Gabriel Thomkins,
William Winn,
Joseph Laine,
Benjamin Lisk,
John Lisk,
Stephen Weekes,
Abraham Maley,
Matthew Cure,
Samuel Cure,
Matthew Buis,
Nicholas Storm,
Peter Storm,
Gessom Bounds,
William Goslin,
Abraham Nefuss,
George Nefuss,
Isaac Giou,
John Wood,
Abraham Young,
John Aulgelt,
Thomas Swartwout,
Marvin Rowland,
Thomas Wood,
Joseph Post,
Samuel Kichim,

Nath'l. Gildersleeve,
John Carey, Sr.,
Abr'm Van Hyning,
Ambrose Lating,
Abraham Gerrison,
Abraham Purdy,
John J. Wood,
John J. Schouten,
Charles Venson,
Henry Schouten,
Mar. J. Van Vlaren,
Reuben Gerroson,
John Peck,
Isaac Lecore,
Simeon Mabee,
Lawrence Daily,
Abraham Travas,
John Caunef,
Sutten Bailey,
Isaac Wasbourn,
John Carey, Jr.,
Jeremiah Hett,
Benjamin Hasbrouck,
John Sloot,
Ephraim Scouten,
Henry Light,
Samuel Brown.

Matthias Lyster's Company:

Mat. Lyster, Captain,
A. Herremans, Lieut.,
A. Vanderbilt, Ensign,
John Cook,
John Thorn,
Andrew Burck,
Stephen Thorn,
Hendrick Van Vleck,
Adrian Manfort,

Peter Hoff,
Andr. Herremans, Jr.,
John Herremans,
Stephen Bancker,
John Kennif,
Joshua Besship,
Jacob Johan Dubois,
Cornelius Nostrand,
Abraham Hogeland,

John Hudson,
Dirck Lyster,
John Churchill,
James Hicks,
Francis Brogardus,
Albert Monfoort,
Cornelius Lyster,
Stephen Duryee,
Abraham Lent,
Gideon Tichout,
Johannes Dubois,
John Huff,
John Carnell,
Gerret Lyster,
Abraham Duryee,
Cornel. Van Sickler,

Samuel Livingston,
James Morgan,
Thomas Vanbrare,
Joseph Theale,
Undrel Strong,
Gilbert Strong,
Gilbert Barnes,
Walter Huson,
John Buchout,
John Ses,
John Haboun,
Peter Van Cramer,
Will. H. Harremans,
John Maufoort,
Timothy Sornes.

Captain Horton's Company:

John Brevoort,
John Van Vlaeron,
Adam Aulgett,
Joseph Brush,
John Snedeker,
John Weel,
Edward Hougen,
Isaac Lent,
Gerrardus Vermilyer,
David Vermilyer,
Charles McCrade,
Paule Hoff,
Jacob Lewis,

John Wiltsee,
Peter Delany,
Joshua Duly,
Peter Depue,
Benjamin Clapp,
William Juell,
John Clapp,
Abraham Depue,
John Wilddey,
Jacob Jewill,
Abraham Huff,
Thomas Clapp.

Captain Griffin's Company:

Caleb Bishop,
Matthew Obriant,
Benjamin Thurston,
John Churchell,
Thomas Griffin,
Daniel Ward,

George Nostraind,
Henry Underwood,
Henry Van Tessel, Jr.,
Philip Miller,
Joshua Purdy,
James Ward,

Joseph Anderson,
Henry C. Philps,
John Jay,
Benjamin Ackerly,
Solomon Woods,
Andrew T. Schouten,

Benjamin Bloom,
Peter Dubois,
Adrian Covenhoven,
Joseph Thurston,
Philip Verplanck,
Jacob Ward.

By order of the Committee,
DIRCK G. BRINCKERHOFE, Chairman.

Fishkill, August 23, 1775.

Sir:—Enclosed is the return of the persons who have signed the Association, and of those who have refused. In the latter you find many erasures, occasioned by their signing afterwards. This affair has been delayed thus long, on account of pursuing lenient measures.

I am, by order of the Committee, your most obedient servant,
DIRCK G. BRINCKERHOFF, Chairman.

Signers in Beekman's Precinct, Dutchess County, July, 1775.

William Humfrey,
Joshua Carmen,
Ebenezer Cary,
Charless Platt,
William McNeal,
William Clark,
Thomas Ley,
Samuel Crandel,
Maurice Pleas,
Thomas Nethaway,
Benoni Sweet,
Nathaniel Stevenson,
Nathaniel Cary,
Samuel Lewis,
Zebulon Ross,
Samuel Gardiner,
Martin Cornell,
Benjamin Noxon,
Elial Youmans,

John Forguson,
Henry Whikmon,
Nuklus Omey,
Walton Huling,
John Huling,
Jacob Miller,
William McDowell,
Thomas Cornell,
Isaac Dennis,
James Humfrey,
Thomas Spencer,
William Bently, Jr.,
Fr. West,
John Jenkins,
Aholyab Markes,
Arnold Reynolds,
Amos Randall,
John Wightman,
Whiten Parkes,

Jonathan Dennis,
Gideon Hall,
Jabez Spencer,
John Eagles,
John Sweet,
James Wells,
Job Shearman,
Joseph Carr,
Daniel Uhl,
William Smith,
Samuel Sweet,
Peter Shear,
Peter Shear, Jr.,
Roger Mory, Jr.,
Isaac Yerrington,
Peter Storm,
Josiah Ingersol,
James Mc Lees,
Nathaniel Wicks,
John Weaver,
Edward Howard,
William Hall,
Joseph Carr,
Joshua Champlies, Jr.,
Isaac Vail,
John Arnold,
Job Tanner,
Johannes Delong,
Hezekiah Rogers,
Ezekiel Rogers,
Griffin Reynolds,
Peter Brill,
Samuel Cornwell,
Josep Lawless, Jr.,
Peter McClus,
John Hopim,
Zephaniah Brown,
Cornelius Van Wyck,
Joshua Carman, Jr.,

John Melony,
John Andrews,
Charles Newton,
Henry Bailey,
Francis Losee,
Daniel Smith,
William Shear,
William Champlin,
Philip Vincent,
John Vinton,
Stephen Forgoson,
Jonathan West,
John Kelly,
Benjamin Fargason,
Joseph Reynolds,
Maurice Smith,
Joseph Taylor,
Steven Johnson,
James McCollom,
Edward Weaver,
Gershom Thorn
Peter Harris,
William Brewer,
James M. Creedy,
Abraham Hyatt,
Gilbert Totten,
Edward Tredwell,
Elias Alley,
Isaac Calton,
Peter Harris,
James Vosburgh,
Jesse Oakley,
Tillinghast Bentley,
Peter Noxon,
Thomas Doxsle,
Henry Pearsall,
Garret Mill,
Johannes Lain,
Henry Smith,

REVOLUTIONARY PLEDGE. 115

Lodovick Sweet,
George Sweet,
David Storm,
Salmag. Edwards,
Stephen Townsend,
Joshua Burch,
David Brill,
Nicholas Koons,
Benjamin Birdsall,
Christopher Wait,
David Sweet,
John Moon,
Nicholas Potter,
Judiah Jenkins, Jr.,
Jonathan Jenkins,
Thomas Clark,
John Hill,
Andrew Cockrane,
Timothy Force,
Clear Everit,
Ezekiel Smith,
Benjamin J. Rish,
Isaac J. Rish,
Rowland Stafford,
William Bentley,
Tabor Bentley,
Thomas Baker,
William Spencer,
John Bentley,
Nial Tripp,
Daniel Fish,
Judiah J. Rish,
Solomon Force,
Benjamin Force,
Seth Sprague,
Benjamin Spencer,
Samuel Whitman,
Matthew Coon,
Nathaniel Sweet,

Casy Eldridge, Jr.,
Johannes Lossing,
Samuel Tomson,
Benjamin Hall,
Abel Parker,
James Tanner,
Joshua Champlin,
Benjamin Force,
Abraham Denne,
Joseph Denne,
Richard Mackrill,
Jacob Lain,
John Beam,
Henry Shear,
Theophilus Sweet,
John Wooley,
William Tanor,
Charles Heayelton,
John Snider,
Seth Smith,
Jacob Esmond,
John Sweet,
Elisha Champlin,
Joseph Holloway,
Jacob Hutchins, Jr.,
John Oats,
James Eastmond,
Lewis Shear,
Israel Vail,
David Storm,
Jonathan Jenkins,
Gideon Hall,
Ezekiel Hubbard,
Joseph Booler,
John Sweet,
Joshua Mowry,
Stephen Mowry,
Cornelius Meynard,
Tobias Clements,

Nathaniel Rogers,
Andrew Carman,
Albert Adriance,
James Wiltse,
Samuel Young,
Daniel Lawrence,
William B. Alger,
Job Green,
William Humfrey, Jr.,
Joseph Carman,
John Hegerman,
George Losee,
Johannes Acker,
France Wiltse,
Henry Cornell,
Abel Simson,
Zachariah Flagler,
John Reasover,
John Losee,
William Kelley,
William Barber,
Nathaniel Smith,
Caleb Townsend,
Myndert Harris,
Obadiah Cooper, Jr.,
John Hicks,
Peter Leavens,
Joel Edget,
Peter Cartwright,
George Croukhill,
Jonathan Parks,
John Fish,
Woos Dakin,
Digmus Kimee,
John Comptor,
John Lamb,
Jacob Rouse,
Elijah Forgason,
Elijah Forgason, Jr.,
Job Conger,
David Pamer,
David Abbet,
Matthew Beckwith,
Abraham Mosher,
David Cash,
Amos Crandell,
Pardon Fish,
Sylvanus Cash,
Thomas Bullock,
Henry Birdsall,
Nathaniel Sol,
Ebenezer Sol,
David Brown,
Samuel Euery,
Addom Bockus,
Nehemiah Lester,
Jonathan Alger.

The following are the names of those persons who refuse to sign the Association of *Beekman's* Precinct, Dutchess County:

Arey Delong,
James Gaslin,
Peter Rossell,
Jacob Hasver,
Matthias Valentine,
Richard Heliker,
William Harris,
Richard Tripp,
Richard Tripp, Jr.,
Israel Tripp,
James Noxon,
Barthol. Noxon, Jr.,
Michel Woolf,
Smighting Tripp,

Peter Hogoboom,
Daniel Beadle,
John Wilkenson,
Christopher Moyer,
Myndert Valey,
Henry Gidley,
John McDonald,
Samuel Smith,
Martine Easterly,
Daniel Ferris,
James Burtice,
Nathan Hyatt,
Frederick Shapher,
Thomas Brundage,
Peter Levins, Sr.,
William Bocker,
Baultis Veily,
Bartholomew Wood,
Abraham Byce, Jr.,
Peter Chatterton,
Philip Miller,
Lawrance Lossee,
Israel Titus,
John Brown,
Robert Thorn,
Stephen Lockwood,
Peter Paley,
Jonathan Thorn,
Peter Dop,
Peter Johnson,
Johannes Miller,
Jeremiah Leuderbeck,
Philip Flagler,
William Giles,
Daniel Way,
John Smith,
Garret Burtis,
Martine Overaker,
Cornberry Dayton,

Myndert Cole,
Josiah Bull, Jr.,
Charles Thomas,
Gilbert Thorn,
John Akerbry,
Cornelius Hegeman,
Jonathan Atherton,
William Woolf,
Aaron Lasey,
Crapo Lake,
Francis Delong,
John Burnit,
Stephen Dean,
Samuel Stringham,
Ichabod Bourman,
Sylvester Richmond,
James Titus,
Ephraim Horton,
Edward Adams,
Thomas Hutchings,
Robert Moon,
James Striker,
Ebenezer Worden,
Charles Vincent,
William Sleeves,
Thomas Langdon,
Peter Buyce, Jr.,
Samuel Emory,
Rowland Emory,
Jacob Brill,
Jeremiah Haxstum,
Elias Palmer,
Benjamin Kenyon,
Nicholas Mosher,
Richard Cornell,
Peter Deeyo,
James Pettet,
William Gifford, Jr.,
Capt Yerry Emigh,

Peter Simson,
Lawrence Emigh,
Samuel Whipple,
Isaac Veal,
Philip Emigh,
Nicholas Emigh, son of
    Philip,
Hendrick Emigh,
John Ball,
Hendrick Klyn,
John Dearstine,
Abijah Ketcham,
Michal Shearman,
Amos Pine,
Nathan Hoag,
Peter Emigh,
Richardus Cornell,
Valentine Stover,
Richard Vincent,
Preserved Fish,
Joseph Losce,
Capt Joseph Harris,

Lieut. Hey. Collins,
Ensign Barnt Veily,
Abraham Buyce,
Causper Overhiser,
William Gifford,
Roger Morey,
Samuel Crandle,
Samuel Crandle, Jr.,
Peter Kedney,
Oliver Waterman,
Jesse Thorn,
Jacob Ferguson,
Johannes Shear,
Charles Davis,
Jasper Fullmore,
Andrew Skidmore,
John Golder,
Capt. Michael Vincent
Lieut. Peter Buyce,
Ensign Steph. Hunt,
Yerry Lossing.

Signers in Poughkeepsie, Dutchess County, June and July, 1775:

Zepaniah Platt,
Peter Tappen,
Samuel Dodge,
William Forman,
John Baily, Jr.,
Johannes Swartwort,
Bicter Van Kleeck,
John Freer,
Henry Livingston, Jr.,
Elias V. Van Bunschoten,
Robert North,
Lewis Dubois,
Andrew Billings,
Peter Low,

Ezekial Cooper,
John Schenck, Jr.,
Paul Schenck,
Jacobus Freer,
John Romyne,
Andrew Wattles,
Nathan Tray,
Barent Lewis,
Thomas Holmes,
Jacob Van Bunschoten,
Abraham Fort,
Carel Hoefman,
Henry Hoff,
Gorus Storm,

Thomas Jacockes,
Barnardus Swartwort,
Francis Jaycock,
M. Van Keuren,
Azariah Winchester,
Henry Willsie,
John Willsie,
William Sawckes,
Thomas Burnet,
James Brisby
Matthew Burnett,
Gideon Boyse,
Thomas Bont,
William Lawson, Jr.,
Abr'm Van Keuren,
John Saunders,
John Briener,
Hans Berner,'
Benjamin Jaycock,
Thomas Rowse,
Isaac Poole,
Jonathan Johnson,
Aaron Reed,
John Pilgrit,
Peter Lossing,
Peter Horn,
William Burnett,
James Elderkin,
John Waterman,
Johannes Fort,
Simon W. Lossing,
Mat. Van Keuren, Jr.,
Silvanus Greatwaks,
Samuel Smith,
James Livingston,
Richard Davis,
Law. Van Kleek,
John Mott,
Richard V. Denbergh,

Simon Freer,
John Davis,
Robert Noa,
Isaah Bartly,
John Schenck, Jr.,
Hendrick Pells,
Hendrick Pells, Jr.,
Johannes Kidney,
Jacobus Schryver,
Henry Hegeman,
George Sands,
Hobert Waddel,
Myndert Van Kleek,
Henry Ellis,
Henry Van Blercome,
Simon Leroy,
Henry Kip,
Benoni Kip,
Abraham Banlay,
M. Van Denbogart,
Isaac Kornine, Jr.,
Alexander Grigs,
Simon Bartley,
Peter Tappen,
Robert North,
Ezekiel Cooper,
William Terry,
Alexander Haire,
Thomas Poole,
Tennis Tappen,
Nathaniel Hemsted,
George Brooks,
Nathaniel Conklin,
John Townsend,
Andrew Billings,
Samuel Corey
John Tappen,
Kenry Dodge,
Jonas Kelly,

Stephen Hendrickson,
Nathaniel Ashford,
Andrew Weeks,
John Ter Bush,
Cornelius Noble,
James Brisleen,
John Johnson,
Abraham Pitt,
Samuel Cooke,
James Winans,
John Seabury,
William Forman,
Henry Livingston,
S. Van Voorhees,
John Conkling,
Matthew Conkling,
Thomas Travis,
Zachariah Burwell,
Lemuel Howell,
Abraham Swartwout,
Richard Everitt,
Matthias Sharp,
John C. Hill,
John T. Van Kleeck,
Dorthir Conner, Jr.,
James Read,
Richard Warner,
William Kelly,
James Lewis,
George Shannan,
Albo. Watervell,
William Roach,
Elias Freer,
Leonard Van Kleeck,
Richard Snedeker,
F. Van Denbogart,
Gerrit Van Wagenen,
Jac. Van Kleeck,
Henry Eliss,
John Maxfield,
L. J. Van Kleeck,
Lewis Dubois,
Jacobus Frear,
John Reed,
Jacob Rhoades,
William Wilsey,
Michel Yerry,
Ephraim Adams,
Tunis Hannes,
Matthew Dubois,
E. Van Bunscoten,
Martin Bush,
Hendrick Bush,
James Luckey,
Samuel Luckey,
Abraham Ferden,
Peter F. Valleau.
Wilhelmus Ploegh,
Geleyn Ackerman,
Joel Dubois,
Peter Mullin,
Simon Leroy, Jr.,
David Dutcher,
Peter Van Dewater,
Edward Symmonds,
Cornelius Viele,
Eli Read,
Peter Low,
Larrine Lossing, Jr.,
John Dubois,
Casparus Westervelt,
Lodowick Sypher,
Christian Bush,
Silvanus Beckwith,
Alex. Chaucer,
Caleb Carmen, Jr.,
John Van Kleeck,
John Seabury, Jr.,

Joshua Moss,
S. Van Denburgh,
Nathaniel Dubois,
C. R. Westervelt,
Cornelius Westervelt,
Enyamen Westervelt,
C. B. Westervet,
Peter Andes Lansing,
William Annely,
William D. Lawson,
John C. Ringland,
Gerrit Van Vliet,
Jeremiah Dubois.
Jacob V. Denbergh,
Peter Van Vliet,
Robert Hoffman,
William Jones,
Jacob Low,
Bernardus Swartwout,
John L. Van Kleeck.
Minnard Swartwout,
John Swartwout,
Frederick Van Vliet,
John Robinson,
John Bailey, Jr.,
Jac. Van Denbogart.
Caleb Carmen.
Jacob Ferris,
Omar Ferris.

A list of men's names who refused to sign the General Association recommended by the Provincial Convention, taken at Poughkeepsie, June and July, 1775:

James Kelly,
H. Van Denburgh,
H. Van Denburgh,
H. Van Denburgh, Jr.,
Nathaniel Babcock.
Felix Lewis,
Austin Crud,
Tunis Williamson.
B. Noxen,
B. Crannell,
Melancton Lewis,
Peter Dubois, Jr.,
John Ferdon,
Zachary Ferdon,
Jacob Ferdon,
Esquire Ferdon,
John Miller,
Arie Medlar,
William Lassing,
Samuel Hull,
Isaac J. Lassing,
Flemming Steenbergh.
George Ame,
Jonathan Morey,
Samuel Pinckney.
Myndert Kidney,
Jacobus Kidney,
Jeremiah Dubois,
Evert Pelts,
Francs Pelts,
Michel Pelts,
Nehemiah Veal,
Jacob Polmatier,
Robert Kidney,
Abraham Frair,
Abraham Frair, Jr.,
Matthew Kipp,
Simon Frair, Jr.,
John Bomen,
Michael Wellding,
John V. D. Bogart,
Joseph Chaddirdon,

John Hunt,
James Latsing,
Myndert Byndirs,
Eli Emons,
John Emons,
John De Graff,
Baltis Van Bleek,
Matthew Kipp,
James Wood,
James Douglass,
Aaron Olmstead,
Henry Beyex,
Eli Read,
Ebenezer Badger,
Peter P. Van Kleek.
Gail Yelverton,
John Palmitear,
John Coopman,
Thomas Freer,
William Emott,

Michael John Rutsen,
George Baldwin,
Hendrick Miller,
Henry Barnes,
Robert Churchill,
Isaac Baldwin,
Isaac Baldwin, Jr.,
Elias Thompson,
John Van Deburgh,
H. J. Van Deburgh,
Peter Van Deburgh,
William Barns,
Simon Noxen,
John Low,
William Low,
Thomas Pinkney,
Ezekiel Pinkney,
John Pinkney,
Henry Barns,
Peter Laroy.

North-east Precinct, Dutchess County.

Seth Case, Jr.,
Charles Graham,
Benaniwell Denel,
David Harvey,
Seth Case,
Thomas Merit,
Icabod Case,
James Hodges,
John Bull,
Stephen Trusdell,
Benjamin Egelston,
Jonathan Lawrence,
Luther Holly,
John Porter,
Joshua Hamblin,
Elisha Colver,
Archibald Johnston,

Samuel Nooly,
Simon Dakin,
Ebenezer Hartwell,
Josiah Holly,
Seth Perry,
David Lawrence,
Ebenezer King,
Abraham Hartwell,
Gilbert Clapp,
Joseph Rundel,
Jeremiah Brownel,
Uriah Lawrence,
James Atwater,
Philip Spencer,
Joseph Peck,
Samuel Roe,
Stephen Merritt,

Alex. McMullin,
John Buttolph,
Adam Stevens,
Thomas Knapp.

A true return of the names of those that refused to sign, given by me.

URIAH LAWRENCE,
DAVID BOTOLPH,
PETER KNAPP,
JOHN HALLEY.

Northeast Precinct, Dutchess County.

Samuel Kie,
Hugh Rea,
Elisha Mead,
Robert Orr,
John Orr,
Jehiel Mead,
Joseph Loggan,
William Smile,
John Crandle,
Hugh Orr,
Daniel Wilson,
Samuel Mott,
Ebenezer Young,
David Love,
Daniel Parks,
David Hamblen,
Peter Knickerbacker, Sen.,
L. Knickerbacker,
P. Knickerbacker, Jr.,
J. Knickerbacker,
Robert Wilson,
James Wilson, Jr.,
John Wilson,
John Carey,
Gulman Alitzer,
Matthew Orr,
William Rea,
Joseph Foster,
Jesse Ferris,
Wintrip Norton,
Joseph Palmer, Jr.,
Johnynal Meton,
James Headding,
Silence Jackson,
Seth Fish,
Isaac Winans,
Jeremiah Giffers,
James Wilson, Sen.,
Frederick Stickels,
John Link,
John Fulton,
John Rouse,
Edward Edsed,
Benjamin Soule,
John May,
J. Salisbury, Sen.,
David Bostwick,
William Parks,
John Bortell,
Stephen Edgaat,
John Avery,
George Edgeet, Jr.,
Jonathan Smith,
John Horn,
Samuel Crandell,
William Robbins,
Peleg Horten,
Michal Masfield,
Moses Fish,
John Carpenter,

Asahel Owemer,
Elijah Lake,
Barnt Van Kleck,
Oliver Evans,
Joseph Palmer.

A list of the persons that refused to sign this *Association:*

William Clum,
Philip Clum,
Jonathan Batreck,
William Batreck,
Jacob Loucks, Jr.,
Peter Allen,
Isaac Allen,
Jacob Drum,
Zechri Tetr,
Nicholas Row,
John Hipman,
John Drum, Jr.,
John Houk,
John Row,
John Row,
Peter Row,
John Kristr,
George Miner,
John Drum,
Zechri Philips,
John Backes,
Yerre Keffr,
Martis Kreepr,
Frederick Destr,
Jacob Row,
Peter Bitchr,
Adam Bitchr,
Andres Houk,
Peter Bosson,
Honesfelt Shaw,
Simon Killmore, Sen.,
Jacob Killmore,
Wynat Weever,
Honthise Couse,
John Houghtaling,
Jacob Hover,
Andrew Collson,
John White,
Joseyh Mott,
William Green,
Nehemiah Avery,
Amos Avery,
Michal Coloney,
Daniel Mead,
Elisha Davis,
William Davis,
William Davis,
Peter Couse,
Jacob Houghtaling,
Zost Hendrick,
Wise Row,
Derick Fendick,
Frederick Horn,
Elijah Forgason,
Jeremiah Forgason,
Ruban Crandell,
John Philips,
Gerret Holsop,
Frederick Stickle,
John Link,
Jacob Shaver.

Dutchess County, Northeast Precinct, July 5, 1775.

The foregoing is a true return of the names of the Inhabitants and Freeholders in the District allotted to us, that signed this *Association*, and the names of those that refused to sign this *Association*.

P. KNICKERERBACKER, DANIEL WILSON,
HUGH ORR, J. REISENBERGER, JR.

Dutchess County, Northeast Precinct.

Ebenezer Bishop, Ebenezer Crane, Jr.,
Levi Stalker, Philip Lott,
Cornelius Fuller, Charles Trupell,
David Bulkley, Wheaton Robinson,
Thomas Crosby, Ebenezer Merrit,
Joseph Jackson, George Morhouse,
David St. John, Levi Rawlee,
Thomas Crosby, Jr., James Winchell,
Renel Seton, Jonathan Grenell,
Willard Seton, Joseph Stalker,
Benjamin Crosby, Ebenezer Crane,
John Seton, Thomas Townsend,
Comfort Stalker, Benjamin Covey,
Vincent Foster, James Coval,
John Wilkie, Caleb Woodard.

Dutchess County, Nine Partners, }
Northeast Precinct, July 5, 1775. }

The above and foregoing is a true return of the names that were willing to sign this Association; and the names of those in the District that refused are on the other side of this Association paper.

GEO. MORHOUSE, per Sub-Committee.

The list of Persons not signers:

John McAlpine, Daniel McAlpine,
Walter McAlpine, McQuin, a young man lately
Darby Lindsey, from Scotland,
Lewis Bryan, James Bryan.

Dutchess County, Northeast Precinct.

Silas Husted, Henry Wiltse,
Morris Graham, Henry Sherburne,

11*

Gideon Salsbury,
Augustin Graham,
John Shirar,
John Colvin,
David Orr,
John Colupland,
John Hayes,
Asa Bullock,
William Orr,
Daniel Palmer,
Samuel Crandell,
Samuel Crandell,
John Row,
John Brown,
Israel Thompson,
Richard Estes,
John Burnet,
John Sa,
Samuel Couger,
Orra Forgoson,
John Catten,
William Stewart,
James Ralstan,
John Head,
Edward Senary,
Lemuel Winchel,
George Head,
Bernard Ostrim,
James Alit,
John Melham,
Benjamin Southard,
Benjamin Cuthbert,
J. Simmons,
George Schneyder,
Cornese Dekmettac,

Smith Simmons,
Robert Enery,
Cornelius Wels,
Casper Rowe,
Simon Gifford,
Nathaniel Mead,
Jonathan Mead,
Kemuel Leed,
Simon G. Myer,
Lemuel Williams,
John Crandell,
Benjamin Congar,
Cornelius McDanniel,
John Crandell,
Joseph Crandell,
Phineas Rice,
James Stephens,
James Newcomb,
Adonijah Newcomb,
John Lennon,
Samuel Miller,
James Winchel,
Andrew Quick,
Aaron Darling,
Isaac Lamb,
Bostion Row,
Wm. H. C. Derry,
Claudius Delis,
George Robertson.
Caleb Norton,
Asa Bishop,
Ensley Simmons,
Garner Stuart,
John Williams,
John Hoff.

Northeast Precinct, Dutchess County, }
July 5, 1775. }

A true return of the names of the Inhabitants of the several Districts allotted to us to hand about this Association.

WILLIAM STUART, MATTHEW MEAD,
J. SIMMONS, FREDERIC HAM.

List of persons who refused to sign :

Nicholas Silvernail,
George Hookingham,
Oliver Asten,
Elisa Colvin,
Nathaniel Niles,
Abraham Osstrander,
John Van Ramp,
Jacob Brinstool,
Thomas Gray,
Henry Teets,
Asa Brown,
Jacob Donehen,
Tenes Teelen,
Abraham Scouten,
Coonrad Melham,
Jacob Van Bramer,
John Smith,_
Christopher Teal,
John Merrehew,
Robert Embray,
Philip Easter,
John Pitchor,
George Martin,
George Shoemaker,
Aaron Shaw,
Daniel North,
Casper Bell,
Matthew Winter,
John Wilde,
Richard Wilde,
William Wibs,
Obadiah Gefford,
William Stuart,
J. Simmons,
Nathaniel Meade.

Dutchess County, Northeast Precinct.

Joseph Ketchum,
Joseph Ketchum, Jr.,
Jonathan Mapes,
Alden Ashley,
Benjamin Perry,
Josiah Perry,
William Hager,
Richard Denton,
Samuel Egelston, Sr.,
Samuel Denton,
Samuel Egelston, Jr.,
Ephraim Jones,
Seth Calkin,
Hezekiah Ketchum,
Moses Calkin,
Joshua Hamblin,
Joshua Dakin,
Jonathan Dolph,
Josias Denton,
Arsthoe Vancry,
Elijah Calkin,
Jared Carter,

Nathan Attwood,
Isaac Rogers,
Joseph Reynolds, Jr.,
Jonathan Close,
Joseph Rogers,
Abner Wilcox,
Ebenezer Beatch,
David Calkin,
Charles Haw,
Josiah Wilcox,
Lebbens How,
Daniel Baker,
Nathaniel Lothrop.

Northeast Precinct, Dutchess County, July 5, 1776.

A true return of the names of the Inhabitants and the Freeholders in the Districts appointed for me to hand about this Association.

JOSEPH KETCHUM.

Dutchess County, Amenia Precinct, June and July, 1775.

Simeon Cook,
Ichabod Paine,
William Barker,
Job Mead,
Jonathan Shepherd,
Elijah Holmes,
Israel Shepherd,
Abner Gillet,
Jacob Power,
Barnabas Paine, Jr.,
Noah Hopkins,
Elias Besse,
Ichabod Paine, Jr.,
Simeon Cook, Jr.,
James Hebbard,
Samuel Shepherd, Jr.,
David Bruster,
Elihu Paine,
Asahel Sherwood,
John Brusan, Jr.,
Elijah Daily,
Thomas Cornwell,
David Gillet,
Ebenezer Mays,
David Rundel,
Thorn Putney,
Solomon Wheeler,
Thomas Morey,
James Palmer,
Elijah Smith,
Nehemiah Dunham,
Gardner Gillet,
Barnabas Paine,
Joseph Backus,
Elnathan Spalding,
Levi Atwater,
Benjamin Doty,
Benjamin Atwater,
Elijah Porter,
John Atwater,
Ezra Thurston,
Archibald Farr,
King Mead,
Seth Wheeler,
Robert Wood,
Zadock Buck,
Timothy Tilson,
Jacob Spuer,

John Osborne,
John Mead,
Crover Buel, Jr.,
Barnabas Cole,
Jonathan Allerton,
James Barker,
Noah Wheeler,
Daniel Garnsey,
Samuel King, Jr.,
Benjamin Brown,
Matthew Stevens,
William Finch,
Joseph Smith,
Thomas Lawrence,
Ebenezer Carter,
James Alsworth, Jr.,
Barzaleel Rudd,
Rufus Herrick,
Brinton Paine,
Judah Burton,
James Betts,
Beniamin holmes,
John McNeil,
Samuel Herrick,
Benjamin Herrick, Jr.,
William Herrick,
John Curry,
Shubal Tyler,
Samuel Dodge,
Thomas Welch,
Stephen Herrick, Jr.,
Squire Davis,
Abel Hebbard,
Elisha Adams,
Ebenezer Latimore,
Ichabod Holmes,
Samuel Waters,
Justus Wilson,
Wm. Wynants, Jr.

Benjamin Crofoot,
Benjamin Denton, Jr.,
Joel Denton,
Benjamin Denton,
Jacob Reynolds,
James Beadle,
Benjamin Fowler,
William Knapp,
Abner Holmes,
Nathan Herrick,
Isaiah Mead,
Theoph. Lockwood,
Levi Mayhew,
John Howard,
William Ford,
Jesse Kinne,
Daniel Shepherd,
Roswell Hopkins,
Samuel King,
Abraham Paine,
John Brunson,
Jonathan Buck,
David Collin,
Zebulon Rudd,
Peter Morse,
Paul Johnson,
Nathan Spuer,
Israel Buck,
John Thayer,
Joseph De Lavergue,
Even Jones,
Joab Cook,
Jesse Smith, Jr.,
Enock Crosby,
John Mordach,
Ebenezer park,
William King,
Grover Bull,
Isaac Parks,

Parrock Sherwood,
William Cornwell,
Samuel Cornwell,
Lewis De Lavergue,
Thomas Smith,
Gabriel Dickson,
Timothy Green,
John Holms,
Ezekiel Johnson,
William Alsworth,
John Denney, Jr.,
William Wilsey,
John Bartow,
Elijah Roe,
Isaac Marks,
James Barnet,
Gideon Castte,
Nathaniel Cook,
Benjamin Vaun,
Samuel Holmes,
Stephen Hinne,
Jabez Crippin,
Lawrence Wiltse,
Joseph Fowler,
John Denton,
Abraham Adams,
Isaac Burton,
Daniel Blaksly,
Robert Wilson,
Joel Ketchum,
Ebenezer Kinne,
Richard Brush,
Benjamin Herrick,
Edmond Perlee,
William Blunt,
Monmouth Purdy,
Jacob Elliot,
Stephen Reynolds,
Joshua Talcut,

Ezra Cleavland,
Samuel Thompson,
John Coy,
Stephen Herrick,
James Smith, Jr.,
Beriah Thomas,
Isaac Burton, Jr.,
Mayhew Dogget, Jr.,
Nathaniel Foster,
John Drake,
David Brown,
William Moulton,
Ezra Bryan,
James Allen,
Eli Burton,
Sam'l Thompson, Jr.,
John Ford,
John Thurston,
William McCollough,
Jonathan Fish,
John Farr,
John Douglass,
Joest Power,
Elijah Wood,
Reuben Wilson,
Daniel May,
Moses Harris, Jr.,
William Reynolds,
John Barnet, Jr,
James Ford,
John Jones,
William Adams,
Ephraim Ford,
Abraham Adams, Jr.,
Weight Milleman,
Daniel Davison,
James Dickson,
Elisha Latimore,
John Collins,

John Benedict,	David Waters,
Versal Dickinson,	Lemuel Brush,
William Brush,	Jason Hammond,
Platt Smith,	David Trusdel,
Josiah Webb,	Job Milk,
Sylvester Handley,	Adin Tubbs,
Elijah Kinne,	Jared Rundel,
Samuel Benedict,	Joel H. Thurston.
John Barnet,	

I do agree to the above Association, so far that it doth not interfere with the oath of my office, nor my allegiance to the King.
ISAAC SMITH.

Not to infringe on my oaths.
ABRAHAM BECKER.

June 8, 1775.

This may certify, to all people whom it may concern, that I, the subscriber, am willing to do what is just and right to secure the privileges of America, both civil and sacred, and to follow the advice of our reverend Congress, so far as they do the word of *God* and the example of *Jesus Christ;* and I hope in the grace of *God*, no more will be required. As witness my hand:
JOHN GARNSEY.

The following persons (three Tories) have neglected to sign the *Association: Joel Harvey, Jun., Philip Rowe; John Garnsey* has signed the paper annexed.
ROSWELL HOPKINS,
Amenia, July 12, 1775.

Gentlemen: Agreeable to your request, I have procured the persons within mentioned to subscribe the *Association*, together with Mr. *Samuel King* and Mr. *Silas Marsh*, all in *Amenia Precinct*, in *Dutchess* County. The two lists of Mr. *Marsh* and this have four hundred and twenty signers, and six have delayed or refused.
I am, Gentlemen, yours, &c.,
ROSWELL HOPKINS.

Dutchess County, Amenia Precinct.

Abraham Slocum,
John Mead,
John Freeman,
Joel Washburn,
Nathan Gates,
Thomas Thomas,
John Seymour,
Stephen Warren,
Eleazer Gilson,
James Mead,
Alexander Hewson,
Jared Brace,
Eliakim Reed, Jr.,
Samuel Dunham,
John Torner,
Martin De Lemetter,
Joseph Doty,
Samuel Sniter,
Samuel Jarvis,
Lot Levitt,
John Boyd,
Matthew Vandeusen,
Nathaniel Swift,
Eleazer Morton,
Isaac Osburn,
Jonathan Hunter,
Samuel Swift,
Ashbel Winegar,
Reuben Doty,
William Hunt,
Nicholas Row,
Samuel Gray,
Simeon Reed,
Samuel Southworth,
Elisha Hollifler,
Benjamin Maxam,
Moses Gillett,
Lemuel Shirtliff,
Abial Mott,

Samuel West,
John Cline,
Jehea Rogers,
Robert Freeman,
Joseph Penoyer,
Samuel Johnson,
Jeduthan Gray,
Johabod Rogers, Jr.,
Elijah Freeman,
Peter Shavelean,
Joseph Doty,
Richard Shavelean,
Solomon Shavelean,
Benjamin Crippin,
David Payne,
Heth Kelly,
Nathaniel Pinney,
Ebenezer Bosse,
Joseph Gray,
Josiah Marsh,
Samuel Palmer,
Obadiah Matthews,
Daniel Sage,
James Chapman,
Daniel Harvey,
Thad. Maning,
Amos Penoyer,
Joseph Gillet,
James B. Rowe,
Abner Shabalier,
Jonas Adams,
Thomas Aily,
David Randle,
Benjamin Sage,
Moses Brown,
John Scott,
Gerardus Gates,
Elkanah Stephens,
John Mears,

REVOLUTIONARY PLEDGE.

Andrew Stephens,
Josiah Cleavland,
John Connor,
Richard Larrabe,
Zedekiah Brown,
Henry Barnes,
Jonah Barnes,
Benjamin Johns,
Ebenezer Larrabe,
Ezra St. John,
Obed Harvey,
Robert Patrick,
Isaac De Lemetter,
Thiel Lamb,
Benjamin Delano,
Daniel Webster,
Samuel Judson,
William Mitchell,
Henry Winegar,
William Young,
John Barry,
James Reed,
John Chamberlain,
Colbe Chamberlain,
Ezra Reed,
Dan. Barry,
David Doty,
John Sackett,
Garret Winegar,
Walter Lothrop,
Ezekiel Sackett,
Increase Child,
Elisha Barlow,
Corns. Atherton,
Reuben Doty,
Sylvanus Nye,
Edmund Bramhall,
Elijah Reed,
Stephen Delano,
Gershom Reed,
Moses Barlow,
Solomon Armstrong,
Thomas Ganong,
Elihu Beard, Jr.,
Nathan Palmer,
John De Lemetter,
William Chamberlain,
Nathan Barlow,
Simeon Hellsy,
Zadock Knapp,
Benjamin Hollister,
John Sackett, Jr.,
Robert Hebard,
Joshua Losel,
John Marchant,
Daniel Castle,
Abraham French,
Seelye Trowbridge,
Asa Foot,
Barnabas Gillet,
Elijah Smith,
John Lloyd,
Epraim Besse,
Robert Johnson,
Jonathan Pike,
Gilbert Willett,
Thomas Mygatt,
Obed Harvey, Jr.,
Silas Roe,
Nathaniel Gates,
Seth Dunham,
Caleb Dakin,
George Sornburgh,
Frederick Sornburgh,
Isaac Darrow,
Joseph Adams,
Conrad Winegar,
Levi Orten,

William Hall,
Robert Freehart,
Peter Klyn,
Ledyard J. Charts,

Isaac Lamb,
Elias Shavilier,
Silas Marsh,
Bower Slason.

Sir :—In pursuance of your order, I have procured the above subscribers (true Whigs), and am, Sir, with great respect, your very humble servant, SILAS MARSH.

Joseph Green,
Simon Whitcomb,
William Roberts,
Albert Finch,
Joseph Benson,
Garret Row,
Nathan Barlow,
Abell Marchant,
Rufus Secton,
Henry Winegar,
Dier Woodworth,
John Benson,
Samuel Winegar,
Daniel Lamb,
John Gates,
Edward Bump,

John Dunham,
Richard Sackett,
Stephen Gates,
Daniel Washburn,
Jacob Dorman,
Seth Swift,
Ellis Briggs,
Samuel Heart,
Elisha Mays,
Joseph Williams,
Silas Reed,
Richard Hamilton,
Judah Swift,
Samuel Dunham, Sr.,
Peter Slason.

The black roll of Tories. Though out of my limits, I am compelled to remind you, Gentlemen, of *James Smith*, Esq., who is notoriously wicked.

Signers in Rhinebeck Precinct, Dutchess County.

Petrus Ten Broeck,
P. G. Livingston,
George Sheldon,
William Beam,
John Van Ness,
Herman Hoffman,
Ananias Cooper,

David Van Ness,
Egbert Benson,
Jacob Hermanse,
Andrias Hermanse,
Peter Hermanse,
Zach. Hoffman, Jr.,
Martine Hoffman,

REVOLUTIONARY PLEDGE.

Zacharias Hoffman,
Abraham Cole,
James Everett,
William Bitcher, Jr.,
Jacob More, Jr.,
Christian Mohr,
Lodowick Ensell,
Isaac Walwork,
Samuel Green,
Peter Traver,
Andrew Simon,
Jacob Fisher,
Samuel Elmendorph,
Zacharias Backer,
Johannes Hannule,
Johannes Richter,
Levi Jones,
Isaac Cole,
Hendrick Miller,
Simon Cool, Jr.,
Frederic Weir,
John Banks,
H. I. Knickerbacker,
William Tuttle,
Stephen Sears,
Joseph Houlsworth,
Jacob Thomas,
Philip Feller,
Harmen Whitbeck,
Evert Vosburgh,
John Moore,
Philip J. Moore,
Nicholas Hoffman,
John Williams,
Joseph Lenercree,
Jacob Vosburg,
James Doglas,
John Garrison,
Nicholas Hermanse,

Philip Bonasteal,
Simon S. Cole,
Andres Michel,
John Lewis,
Christeaun Miller,
William Klum,
Johannes Miller,
Thomas Lewis,
Hendrick Livey,
Everhart Rydders,
Henry Kuneke,
George Sperling,
Elias Hinneon,
Samuel Haines,
Peter Ledewyck,
Jacob Elemendorph,
Jan Elemendorph,
Patt. Hogan,
Evert Hermanse,
John Cole,
Petrus Bitcher,
Zacharias Roob,
John Balist,
Helmes Heermanse,
Cornelius Elmendorph,
Philip Staats,
John Staats,
Peter Staats,
Isaac Beringer, Jr.,
William Waldorn,
Frederick Benner,
John Hermanse,
Stoffle Waldorn,
Johannes Benner,
George Sharpe,
Christeaun Backer,
Petrus Backer,
Johannes Backer,
Coenradt Lescher,

Michael Sheffel,
Goetlieb Mardin,
Hendrick Mardin,
David Martin,
Cornelius Swart,
James Adams,
Daniel Oeden,
Jacob Schermerhorn,
Cornelius Schermorn,
Reyer Heermans,
Jacob Heermans,
William Bitcher,
Wilhelmus Bitcher,
John Hermanse,
Godfrey Gay,
Hendrick Teter, Jr.,
Abraham Teter,
Johannes Smith,
Jacob Meyer,
Edward Wheeler,
Peter Hoffman,
William Beringer,
Conrad Beringer,
Henry Klum, Jr.,
C. Oosterhoudt,
Benjamin Myers,
John Oosterhoudt,
Peter Cole,
Simon Kool,
Jacob Maul,
Everardus Booardee,
Simon Westfall,
Jacob Tremper,
William Radclift,
H. Waldorph, Jr.,
Henrich Benner,
Jacob Moul, Sen.,
Benj. Van Steenburgh,
Johannes Van Keuren,

Tobyes Van Keuren,
John Klum,
Godfrey Hendrick,
Jacob Beringer,
Joseph Younck,
Christian Fero,
Reyer Schermerhorn,
Wilhelmus Smith,
Frederick Moul,
George Reystorf,
William Harrison,
Christoff Schneyd,
Christopher Fitch,
John Schermerhorn,
Henry Waterman, Jr.,
Jacob Waterman,
Henry Litmer,
John Mares,
Isaac Mares,
James Ostrander,
Christopher Wever,
Peter Westfall, Jr.,
Henry Gisselberght,
John Bender,
Zacharias Whiteman,
Joseph Hobart,
William Schultzs,
John Blair,
Thomas Greves,
Michal Schatzel,
Joseph Rogers,
Benjamin Bogardus,
Hans Kierstead,
Isaac Kipp,
Jacob J. Kipp,
Henry Beekman,
Evert V. Wagenen,
Art. V. Wagenen,
Philip Hermanse,

REVOLUTIONARY PLEDGE. 137

W. Van Vredenburgh,
Jacob Kip,
Jacob A. Kip,
John Tremper,
Henry Shop,
Peter Shopf,
Hendrick Moon,
Herick Berrger,
Johannes Turck,
John White, Jr.,
John Cowles,
Herman Duncan,
John Denness,
William Waldrom,
Cornelius Demond,
S. V. Bunscoten,
B. Van Vredenburgh,
Peter Scoot,
Jonathan Scoot,
John Mitchell,
David Mulford,
Lemuel Mulford,
James Lewis,
Peter D. Witt,
John Pawling,
Olbartus Sickner,
Andrew Rowan,
Martines Burger,
Johannes Scott,
Jacob Sickner, Jr.,
Barent V. W genen,
Jacob Sickner,
J. Van Aken,
Peter Van Nauker,
Jacob N. Schriver,
Paul Gruber,
Solomon Powell,
Henry Bull,
George Bull,

William Powell,
Caspar Haberlen,
Thomas Umphry,
Abraham Scott
William Troophage,
Alexander Campbell,
Abraham Kip,
Peter Brown,
Jacob Schultz,
John Hufman,
Henry Freligh, Jr.,
R. Vhoevanburgh,
Peter Radclif,
Simon Schoot, Jr.,
William Schoot, Jr.,
Jacob Lewis,
Jacobus Kip,
William Skepmus,
Johannes P. V. Mood,
William Dillman,
Cornelius Miller,
Simon Millham,
Lawrence Millham.
Jacob Milham,
Simon Milham,
John Weaver, Jr.,
Benj. Oosterhoudt,
Christ. Deninarh,
Abraham Westfall,
John McFort,
William Carney,
Philip Feller, Jr.,
Nicholas Binestale,
Philip Binestale, Jr.,
C. Wenneberger,
Johannes Benner,
Jacob Benner,
Jacob Folant,
John Rogers,

Nicholas Stickle,
Jacob Tell,
John Sater,
John Haass,
William V. Prudenburgh,
Rurif J. Kip,
P. Van Pradenburgh,

Henry Burges, Jr.,
Ulriah Bates,
William McClure,
Joshua Chember,
Zach. Neer,
Nicholas Stickle, Jr.,

Dutchess County, Rhinebeck Precinct.

A return of the names of such persons as have refused to sign the general Association.

EGBERT BENSON
Chairman of the Precinct Committee.

Mordecai Lester,
Peter Prosses,
Timothy Doughty,
Adam Tibble,
Jacob Tibble,
Lodowick Streght,
Peter Em. Schryver,
Peter Freligh,
Steophanus Freligh,
Adam Ecker,
Peter Ecker,
Johannes Ecker,
Adam Jury Ecker,
J. Van Vradenburgh,
Jacob Van Esten,
Zebulon Hallick,
Adam Burgh,
Michael Bruce,
George Stover,
George Anderson,
Zacharias Cramer,
Johannes Cramer,
Johan. Van Esten, Jr.,
Stephanus Burger,
Christian Bargh,
Christian Bargh, Jr.,
John Hallock,

Christian Bruce,
Peter Frusam,
Hendk. A. Schryver,
Marthen Schryvdr,
Marthynes Schryver,
T. Van Benschoten,
E. Van Benschoten,
Egbert Bunchoten,
Harmanus Bunchoten,
John Carnell,
John Sickner,
B. V. Vradenburgh, Jr.,
Henry Pawling,
John Schryver,
David Schryver,
John Brown,
Hendk. Ecker, Jr.,
Jacob Chafer,
John Holmes,
Philip Pinek,
John Pinek,
Philip Pinek, Jr.,
Jacob Elen,
Henry Wederwaks,
Abraham Wederwaks,
Philip Loune,
Bashan Loune,

REVOLUTIONARY PLEDGE.

Anderis Loune,
George Lament,
Jacob Loune,
John Wels, Jr.,
Benjamin Westfall,
Benjamin Wels,
John Dericks,
Jacob Hendericks, Jr.,
John Bander, Jr.,
John Tile,
Joest Schever,
Frederick Schever,
Henry Schever,
Anthony Strant,
Benj. Stienburgh, Jr.,
Hendrick Meyer,
Tunis Boutcher,
Conradt Polver,
Casper Boutcher,
Jacob Yager,
Juery Hoffman,
Nicholas Hoffman,
Johan. Righpenbergh,
Petrus Righpenbergh,
Andris Luych,
Zacharias Drom,
Hendrick Heermans,
Jacobus Kip,
Johan Van Wagoner,
Barent Van Wagoner,
Matthew Van Etter,
Cobus Van Etter,
Isaac Van Etter,
Hendrick Pelts,
Lodowick Elshaver,
Peter Nile,
Coenradt Bammas,
Martha Teel,
Lawrence Teel, Jr.,

Johannes Fraver,
Peter Fradenburgh,
Hans Zipperly,
Jose Neer,
David Lown,
Johannes Lown, Jr.,
Jacob Seeman,
John Seeman,
Jacob Seeman, Jr.,
David Seeman, Jr.,
Jeremiah Seeman, Jr.,
Petrus Fero,
Martin Threecarter,
Bastian Witterwax,
Hendrick Shook,
Christian Shook,
Cobus Shook,
George Shook,
Peter Freligh,
Michael Seeman,
Abraham Seeman,
Jacob Cole,
Jacob Miller,
John J. Cole,
Jacob Shomaker,
George Bennet,
Johannes Sager,
Christian Dederick,
Michael Puls,
David Puls,
Christuffal Puls,
Daniel Puls,
George Puls,
Michael Puls,
Bashan Wagor,
Powlis Wagor,
John Marguet,
Johannes Barker,
Martner Barker,

Lawrence Barker,
George Marguet,
Peter Prough,
Powlis Prough,
Adam Asher,
John Asher,
Gerrit Dedrick,
Jacob Kisel Bargh,
John Kip,
Benj. Van Etten,
Jacobus B. Van Etten,
Jacobus Van Etten,
Jacobus J. Van Etten,
Abraham Van Etten,
Benj. Van Etten, Jr.,
John Van Etten,
Jacob Van Etten,
Philip Traver,
Bastian Traver,
Peter Traver,
John Traver,
Jacobus Vradenburgh,
Jacs. Vradenburgh, Jr.,
Christopher Ring,
George Ring,
Johannes Ring,
David Ring,
Peter Westfall,
John V. Steenburgh,
Gradus Lewis,
John B. Kip,
Hugh Landen,
John Kettyman,
Christian Shults,
John Shults,
Henry Richart,
Dowie Richart,
Philip Richart,
Johannes Richart,

William Wallace,
Henry Wallace,
Francis Nehis,
Charles Nehis,
Francis Nehis, Jr.,
Peter H. Traver,
John H. Traver,
Frederick Traver,
Jacob Traver,
Abraham Kip,
Peter Scriver,
Peter Kip,
Henry Lewis,
Jacob Kelder,
John G. Miller,
William Mackay,
Thomas Briant,
Jacob Smith,
John Tennis,
William Waldrom,
B. Van Benthysen,
Johannes Rysdorf,
Jacob S. Kip,
Cornelius Fynhout,
Corns. Fynhout, Jr.,
Petrus Rysdorf,
Lawrence Rysdorf,
Arent Kipp,
Jacobus Kip, Jr.,
Peter Elkenbergh,
Jacop Evans,
David Shaver,
Jacob Lown,
Peter Van Alen,
Petrus Cram,
Adam Shever,
Jury A. Shufelt,
William Fuller,
Lawrence Shewfelt,

Petrus Shewfelt,
Adam Shewfelt,
John Allemten,
John F. Allemten,
Frederick Slays,
P. Van Benthuysen, Sr.,
J. Van Benthuysen,
Phil. S. Livingston.

Dutchess County, June and July, 1775.

Henry Sherburne,
Jonathan Lewis,
John Hibbird,
Theophilus Wadleigh,
Timothy Soaper,
Samuel Smith,
Daniel Soule,
Jacob Lesh,
Benjamin Atwater.
Titus Mead,
David Robbins,
John Robbins,
Peter Smith,
Jesse Cornell,
Absolom Trowbridge,
Jeremiah Shaw,
Stephen Atwater,
Joseph Crary,
Isaac Smith,
Thomas Hill,
Peter Van Deursen,
Moses Golph,
Ezekiel Kie,
Ira Winans,
Lambert Morey,
Peter Smith, Jr.,
Nathan Lounsbury,
Epentus Lounsbury,
Andus Stickel,
Christian Cambel,
Cornelius Viller,
John Schermerhorn,
B. Knickerbacker, Jr.,
Peter Van Leuven,
Caleb Reynolds,
David Fisk,
Obadiah Holmes,
John Knickerbacker,
Petrus Hommel,
Benj. Knickerbacker,
Caleb Force,
Richard Gray,
Eliphalet Platt,
Isaac Wood,
Phineas Rice, Jr.,
Isaac Young,
James Young,
Jacob Wemer,
Samuel Mabbitt,
Israel Green, Jr.,
Benjamin Terbush,
Gabriel Dowzenbery,
Wilhelm Finche,
Benjamin Crandle,
William Smith,
Motise Wilse,
John Stuart,
Adam Snider,
William Mansfield,
Michael Row, Jr.,
Philip Smith,
John Parkinson,
James Neeson.

July 5, 1775.

We, the subscribers, being duly chosen as a *Sub-Committee*, to return the names of all persons who have signed the above Association; and likewise the persons who did not sign, on the back.

CHARLES GRAHAM,
HENRY SHERBURNE.

A list of the Persons not signers.

John Geo. Kerrick,	John Stickel,
Hontice Smith, Sr.,	John Bearry,
Hontice Smith, Jr.,	Mical Simons,
Nicholas Smith,	Jacob Luke,
Leonard Smith,	Cornelius Clark,
Jonathan Griffin,	Vandil Pulvin,
Jonathan Devall,	John Pulvin,
Tice Wisey,	Hendrick Cufin,
Benjamin Willbor,	Peter Pulvin,
William Merrifield,	Hendrick Hoofman,
Jacob Melions, Jr.,	Philip Snider,
Motise Rowe,	Benj. Vanleuvan,
Daniel McConalep,	Isaac Vanleuvan,
William Melions,	John Weaver,
Lockland McIntosh,	Harry Weaver,
Alexander McIntosh,	Hendrick Row,
William McIntosh,	Giles Weaver,
Andrus Pulvin,	Michael Smith,
William Rector,	Mical Row, Sr.,
Valentine Emert,	John Peter Row,
Hendk. Younklion,	Tuce Smith.

We, the subscribers, inhabitants of the Colony of *New York*, do most solemnly declare, that the claims of the *British Parliament* to bind, at their discretion, the people of the United Colonies in America in all cases whatsoever are, in our opinions, absurd, unjust, and tyrannical; and that the hostile attempts of their fleets and armies to enforce submission to those wicked and ridiculous claims ought to be resisted by arms. And therefore we do engage and associate, under all the ties which we respectively hold sacred, to defend by arms these United Colonies against the said hostile attempts, agreeable to all such laws or regulations as our representatives in the Congress, or future General Assemblies of this Colony, have or shall, for the purpose, make and establish.

SAMUEL WHITTEN.

The names of those who refuse to sign the *Association* in Charlotte Precinct, are :

Peter Hatfield,
Jabez Fineh,
Edward Undrel,
Daniel Sales,
Stephen Hix,
Henry Wecks,
Hendrick Buc,
John Watson,
Edward Mosher,
Matthias Brogue,

Eliphaz Fish,
Joseph Husted,
Richard Simmons,
Jonathan Lapham,
Barnard Hix,
Samuel Titus,
Richard Bartlett,
Samuel Mosher,
Ichabod White,
Uriah Hall.

PHILIPSTOWN.

THIS town by the Act of March 7th, 1788, entitled "an Act for dividing the counties of this State into Towns," is described as follows: "And all that part of the county of Dutchess, bounded southerly by the county of Westchester, westerly by Hudson's River, northerly by the north bounds of the Lands granted to Adolph Philipse, Esq., and easterly by the Long-Lot, number four, formerly belonging to Beverly Robison, shall be, and hereby is erected into a town, by the name of Philipstown." Originally it embraced more than one-third of the county, but has since been diminished by the erection of the town of Putnam Valley in 1839. Its central distance from the city of New York, is about fifty-six miles, and from Albany, ninety-four miles. Its present boundaries are as follows: On the north, by the south line of Dutchess county; on the east, by the west and north lines of Kent and Putnam valley; on the south, by the north line of Westchester; and on the west, extending the whole length of Putnam, by the Hudson river. Its face is broken by hills and mountains, presenting a rough, rugged, and forbidding aspect. Not more than one-fifth of it is under cultivation, and not more than one-third could be made productive, by the most lavish expenditure, of moneys, to the agriculturist.

Let it not be supposed by the reader that it is,

therefore, altogether valueless, although the plough of the husbandman would in vain, and to little profit, be held in its bosom. It contains those materials that are worth more to its owners, than if it was susceptible of the highest agricultural improvement. It is covered with timber, valuable for ship building and other purposes; and, perhaps, from no other township between Albany and New York, for its size, is so great a quantity of wood and timber carried to market. The stone quarries and mineral productions scattered in every direction over its surface, yield a large profit, without any expenditure to the owners of those locations. The burning of charcoal is a profitable pursuit to those engaged in it. The writer has been informed by a farmer owning about 200 acres of land In this town, one-half of which is unfit for cultivation, that during the last year he has realized, from the burning and sale of charcoal alone, $1000 over and above all expenses attending the same.

Although the mountainous and rocky surface of this town will always present an impediment to an extended culture, yet its slopes and valleys in the western part, near the Hudson, are in a good state of cultivation; " and the agriculturist, although he has to labor hard, receives a good return. There are but few men of wealth here, but the inhabitants seem to be in possession of the necessities, if not the comforts of life."

This town, with those now forming the county, as we have before stated, was comprised in the patent granted to Adolph Philipse, by the King of England, and the land let out to those who would come and settle on it, paying no rent for a few years, except the taxes. The tenants, according to the custom of early times, came under the operation of the feudal system, modified, as it had been, by time and that peculiar state of things incident to the settlement of a new and distant country. In most instances they were tenants-at-will; in others for life, or for a certain term of years. This system has exerted its legitimate influence in this town in retarding its agricultural development.

Improvement, at first, proceeded very slowly. Nothing short of actual ownership of land, requiring such large outlays of labor and expense, would stimulate the early settlers to an energetic and extended culture. The tenure by which they held was too uncertain to beget that desire for permanent improvement which stimulates the husbandman when he is sure that the profits of his labor and toil are secure to himself and family. This state of things no longer exists; and nearly all of the land fit for cultivation is now owned by industrious, enterprising men, who have purchased it from the original proprietor, or his descendants. The best farms are found in Pleasant Valley; a vale extending from the Westchester line on the south, to Dutchess on the north, having an average breadth of one mile. It skirts the Hudson until it reaches the village of Cold Spring, then deviates a little to the east, between Bull hill, Breakneck, St. Anthony's Face peaks, on the west, and the central Highlands on the east.

EARLY SETTLEMENT OF PHILIPSTOWN AND PUTNAM VALLEY.

We shall treat of the early settlement of these towns together, as they were originally one; the latter being erected from the former, March 14, 1839. To us such a course seems not only necessary but proper, as the latter has no early history, independent of the former.

With regard to the early settlement of the towns, we have met with more difficulty than with any other article contained in our paper. The aged people are few, whose memories enable them to give dates, with any degree of accuracy, except in a very few instances.

A generation has passed away, who, twenty-five years ago, could have furnished exact chronological information of a very valuable character; but the golden period for collecting it has passed, and we must be content with the imperfect sources that are still left.

The Old Highland Church and Vicinity.—The first settlement in this part of Philipstown, was made by David Hustis, who came from England and settled about half a mile north of the Highland Church—on the road from Cold Spring to Fishkill, and where David Hustis, Esq., now living, resides—in 1730.

He settled down with the Indians around him, and procured the corn, which he first planted, from them. They had about the fourth of an acre under cultivation, the year before, on the east side of the road, a few rods south of the house where the present David now lives. He was the first of the name, and the ancester of the Hustis family, in this town. He became-a tenant-at-will, of the patentee, and rented 310 acres of land, for which he paid a yearly rent of five pounds, or $24, 10. He afterwards occupied 90 acres more west of the first tract, all of which he afterwards purchased. His nearest neighbor was three miles distant, to whom he was compelled to go, a few days after his arrival, to procure fire; his own, from neglect, having gone out.

A short time afterwards, the Haights, Bloomers, and Wilsons, came and settled in the vicinity. At the Highland Church, one Anderson built a house on the site now occupied by the house of S. Birdsall, in which Thomas Davenport, Esq., now resides. A man of the name of Lamoreaux settled there about the same time. Anderson was of Dutch descent, and Lamoreaux, French. Both removed before the Revolution. Benjamin Bloomer was the next settler, who, with one Bush, made a large purchase in the water lot of Roger Morris, between the Church and the Hudson.

David Hustis died in the early part of the Revolution, leaving four sons, Joseph, Caleb, Solomon, and Jonathan. Joseph had three sons, Robert, Joseph, David; and six daughters, Sarah, Abbey, Mary, Charity, Phoebe, Hannah, and died in 1805. All are now deceased, except David and Phoebe.

Caleb had two sons, William and Jonathan; and six daughters, viz.: Elizabeth, Esther, Anna, Rachel, Mary, and Phoebe.

William had four sons, Samuel, Caleb, Isaac, and Josiah, all of whom are now living.

David Hustis was one of the commissioners, who laid out the first roads in the south part of Dutchess County, now Putnam, in 1744.

Cold Spring Village.—Thomas Davenport, great-grandfather to William Davenport, Esq., of Nelsonville, came from England about 1715, and built the first house at Cold Spring. It was burnt down while his son William was living in it.

After his father's decease, William, grandfather to the present William, about 1760, built a small house a few rods distant from where it stood. Isaac Davenport, cousin of William, then built a house on the foundation of the house burned down, which had been built by Thomas, his uncle, and moved the one built by William to it. They now form the old low house on the hill, a few rods distant from the residence of Marvin Wilson, Esq., on the north side of Main Street.

Thomas Davenport had two wives. By the first he had two sons, William and Thomas; and one daughter. By his second he had two sons, and two daughters.

William, grandfather of Willliam now living, was by the first wife, and had one son, named Thomas, who was the father of the present William of Nelsonville, and two danghters, Marybee and Elizabeth. Marybee married Thomas Sutton, Elizabeth married Solomon Cornell, who emigrated to Kentucky. Thomas had one son, William, and two daughters, Sarah and Elizabeth. Sarah married John Snouck, and is dead; Elizabeth married Jonathan Hustis, and is still living. Thomas the great ancestor of the family in this town, died the 30th of December, 1759, aged 77 years. His grandson Thomas, and father of the present William, was born April 11th, 1750, and died ——

Thomas, the brother of the grandfather of the

present William, built a house which stood in the garden just north of the farm-house of Henry De Rhams, on the road leading from Cold Spring to John Garrison's, Esq. Oliver Davenport built the old house, still standing in Nelsonville, opposite to the residence of William Davenport, Esq. Isaac Davenport, the grandfather of Capt. Cornelius and Sylvenus Davenport of Cold Spring, was the son of Thomas Davenport, and born in the old house at H. De Rahams.

The next house was erected by Elijah Davenport, the son of Isaac, and father of Capt. Cornelius and Sylvenus Davenport, and forms the rear of the house now occupied by Asa Truesdell, Esq., on the south side of Main Street. It was built about 1795. Elijah Davenport kept store in it; and in 1817 it was occupied by the Hon. Gov. Kemble, President of the West Point Foundry, as an office. The third house erected in the village, now forms the rear part of the building occupied by Thomas Rogers, at the north end of Foundry Street.

The next house built, in the vicinity of the village, was the old house now standing near Clark's brick-yard, at present unoccupied. It was built shortly after the Revolution by Peter Lindsey, who subsequently sold it to Samuel W. Baird, a cooper by trade, who kept a grocery in it.

An old log-house stood opposite to where William Davenport now resides in Nelsonville, and on the site of the old, low, frame-house now standing. It was occupied by Stephen, brother of the first Thomas Davenport mentioned above.

Thomas Davenport was one of the commissioners for laying out roads in that part of Dutchess County, which is now Putnam, from 1745 to 1755.

Nelson's Highlands and vicinity.—John Rogers made a settlement about 1730, where Cornelius Haight now lives on the old post-road, a few miles north of Continental Village. At that time there was only a path used by the Indians, leading from Westchester through the Highlands, to Fishkill. Having built a log house sufficiently large for a country tavern, he was always sure to have a traveler for his guest during the night, if he reached the house in the middle of the afternoon; as none ever departed on their journey after that time, owing to the danger of traveling through the Highlands after night, and the difficulty of threading such a wild, mountainous, and solitary path. He continued to keep tavern there during the French and Indian war; a short time previous to which, the old post-road was cut through the Highlands by Lord Louden, for conveying his baggage, stores, and troops to the North, to attack the French out-posts. The road followed the Indian path, and is very little altered from the original line. Rogers' was the first house built along the path.

The next house on this road was built by James Stanley, where Samuel Jeffords, deceased, lived about 1750. Thomas Sarles built the next on the east side of the road, between Samuel Jefford's, Esq., and Justus Nelson's, at the foot of the hill, about 1756. The next northward, was built by Elijah Budd, where Joseph Wiltsie now lives, called the Andrew Hill-farm. Gilbert Budd about the same time, built a house where John Griffin, Esq., now resides. Gilbert lived there in the Revolution, and his brother Elijah on the Hill-farm, a quarter of a mile south.

At the south end of Peekskill Hollow, in the now

town of Putnam Valley, a family of the name of Dusenbery and Adams, made a settlement, but at what period of time, we have not been able to ascertain. On the road leading to Peekskill, where Abijah Knapp now lives, George Lane made an early settlement, about fourteen miles east of Continental Village. Nathan Lane settled a little below, where the Hon. A. and S. Smith reside. John Hyatt, who was commissioned a Colonel in the militia in the Revolution, settled between the Lanes.

In Peekskill Hollow, a little south of where Dr. John Tompkins resides, the Post family settled. At what is now called Tompkins Corners, was formerly called the Wickopee and Peekskill Hollow Corners, those roads intersecting each other at that point. Wickopee was the name of an Indian tribe, that lived at Shenandoah, in Dutchess County; and another tribe called Canopus, lived in Westchester, near the line of Putnam, and extending into the hollow which bears their name. Up and down Peekskill Hollow these tribes used to pass, when visiting each other. The lower tribe, when asked by their white neighbors where they were going, when setting out to make one of their visits, would reply, "We're going to see old *Wichopee.*" The name, we believe, is now spelled *Wickapee*, but we have been informed by old people, that it was formerly spelled as above. The Lanes, Posts, Dusenberys, Smiths, and Adams, were original settlers in this town. About two miles south-east of Tompkins' Corners, Abraham Smith, grandfather to the Hon. A. and S. Smith, made a settlement, and purchased a large tract of land from Col. Beverly Robinson, about 1760, where his grandchildren, above-named, now reside.

He emigrated with two of his brothers from England, about 1720, and settled at Smithtown, on Long Island, where they purchased a very large tract. In 1760, he removed to Putnam Valley, and died in 1763. He had one son named Abraham, the father of the Hon. Abraham and Saxton Smith, who reside on the paternal estate. After having surveyed out the tract, he gave the farm, where John and Reuben Barger now live, to one of his chain-bearers for his services.

An early settlement was made by the Ferris family, from New Rochelle, Westchester County. The ancestor of the family came from Rochelle in France, on the repeal of the edict of Nantz, by Louis XIV., in 1685. The family before the Revolution, moved to the vicinity in which Joseph Ferris, now living, resides.

Extract from Town Records.

"At a town meeting in Philipses Precinct in Dutchess county on   day of Apr. 1772 the following officers were Chosen

John Crumpton Clark
Beverly Robinson Supervisor
Joseph Lane
    & } Assessors
Caleb Nelson
William Dusenberry Collector
Israel Taylor
    & } Constables
Isaac Devenport
Justus Nelson
Cor's Tompkins } Poormasters

Cor's Tompkins, Poundmaster for Peekskill Hollow
John Likely Poundmaster for Canopus Hollow
Elijah Budd Poundmaster on the Post Road
Caleb Nelson Poundmaster on the River

Isaac Rhodes & Moses Dusenberry fence Viewers
Isaac Horton & John Jones   do   do
Joseph Haight & James Lamoreaux do   do
Jacob Mandevill & Tho's Devenport do   do

Isaac Rhodes Highwaymaster, for ye Road from Frediraksburg Precinct to the Bridge over Peekskill River near Lewis Jones —

"illiam White Highwaymaster for the Roade from William Dusenberrys up Peekskill Hollow to the Bridge near Lewis Jones which bridge he is to make with his hands & to continue up the Hollow to the Line of Fred'sburgh Precinct—

The remainder of the entry of this Road District, with the next one, of which Robert Oakley was chosen Highway Master, has been nibbled away by mice. The next in order is as follows :

John Winn Highwaymaster for the Road from the Cold Spring Along wycopy Road to the Line of Rumbout's Precinct all the people living north of sd Spring to belong to his Company

Reuben Drake Highwaymaster from Drake's Mills up Canopus Hollow to the Post Road—

John Meeks Highwaymaster on the Post Road from Westchester line to Joseph Bards,—

Elijah Budd Highwaymaster on ye Post Road from Thos Sarles to Rumbout's Precinct

Jacob Mandivell Highwaymaster from the Post Road near Widow Areles through the Highlands to sd Mandivell's House from thence to Caleb Nelson's & from thence to Christopher Fowler's and from thence to the first mentioned Road

John Nelson Highwaymaster from Mr Robinson's Mills to his fathers from thence to Tho's Williamson & from thence to mr Robinson's house

Thomas Devenport Highwaymaster from Caleb Nelsons to his own house & from thence thro the woods to the post Road near Elijah Buds—

Here, again, the mice have destroyed the entry of the remaining road district. After which, is the following entry of a challenge by Uriah Drake, questioning the election of Cornelius Tompkins to the office of Poormaster, and the result on a second ballot.

N. B. all the foregoing persons was chosen Unanimousley Except Cor's Tompkins Poormaster who was opposed by Uriah Drake who demanded a pole at the Close of which

     Cor's Tompkins had 47 Votes
     Uriah Drake   35 do
             12 Defference

upon which Cor's Tompkins was Declared poormaster

April the 25 John Armstrong his mark a Crop of the Right ear

May the 11 1772 John Cavery Desires his mark to Bee Enterd In this Book Which I have Which is a Crop of the neer ear and a Slit in the same and the off ear A Hol and a half Penny and the Half Penny on the under side

May 11 1772 Sibit Cronkit Juneer De Sires his mark to Bee Enterd in this Book Which I have Which is two niks in the neer ear one on Each Side and the off Ear a Slit and a Half Penny upon the under side.

There were thirteen road-districts in this town, then including the town of Putnam Valley, in 1772, and sixteen in 1773. Commissioners for laying out highways were not elected previous to 1773. At the town meeting held in the spring of that year, Joshua Nelson, Moses Dusenberry, and Isaac Rhodes were elected. The record does not contain the names of persons assessed on the different road districts, but only the names of the path masters. The whole number of days assessed on each district were put down on the record against its overseer or master. There is an exception to this, however, in 1783; and it is the only one that we have been able to find upon the earliest record. The entry is as follows:

"a list of amos odell's Company to work the Highway for the present year 1783

	no of Days for Eech man to work
Amas odell	8 Days
John Armstrong to work	8
Jacob armstrong to work	4
William Cristion To work	5
Richard Criston Jun To work	7
Henry Youman To work	4
Oliver odell To work	8
Aaron odell to work ,,	4

The first road laid out by the Commissioners of this town was in 1784; and the description thereof is characterized by more than Spartan brevity. It is as follows:

"May the 10 in the year of 1784 Then we the Comishners Laid out a Road from Calip Nelsons to his Landon Beginin a^t his house Ceepin as near the South of the Brook as near the Brook as Connevent as Can for us

E LIJAH BUDD
HENDRICK POAST
ISAAC RODES

The following is a list of the names which appear on the town record, including Putnam Valley, from 1772 to 1782:

Beverly Robinson,	Israel Taylor,
John Crumpton,	Isaac Devenport,
Joseph Lane,	Justus Nelson,
Caleb Nelson,	Cornelius Tompkins,
William Dusenberry,	John Likely,

PHILIPSTOWN.

Elijah Budd,
Isaac Rhodes,
Isaac Horton,
Joseph Haight,
Jacob Mandevill,
Thomas Devenport,
John Jones,
James Lamoreaux,
Moses Dusenberry,
William White,
John Winn,
Reuben Drake,
John Meeks,
Samuel Warren,
John Nelson,
Uriah Drake,
John Armstrong,
John Cavery,
Sibit Cronkit,
Edward Meeks,
Anthony Field,
Cornelius Gea,
Joseph Nap,
Peter Bell,
Murty Heayerty,
Nathaniel Jager,
Stephen Lawrance,
Jedediah Frost,
Peter Dubois,
Joshua Nelson,
Justus Ones,
Peter Snorck,
Joseph Husted,
John Avery,
Thomas Bassford,
Sylvenus Haight,
Benjamin Rogers,
Stephen Conklin,
Daniel Bugbee,

Daniel Willsie,
John Sherwood,
Reuben Tompkins,
Stephen Devenport,
John Van Amburgh,
Ezekiel Gee,
Samuel Jenkins,
Jacob Reade,
Isaac Odell,
Capt. Israell Knapp,
John Haight,
Hendric Riers,
Amos Odell,
Jacob Armstrong,
William Cristian,
Oliver Odell,
Aaron Odell,
Henry Eltonon,
Robert Oakley,
Thomas Smith,
Joseph Arpels,
William Wright,
Cresterfer Fowller,
Jonathan Ones,
Gabriel Archer,
Sylvenus Lockwood,
Abraham Garrison,
Joshua Mead,
Hendrick Post,
Absalom Nelson,
Peter Ryall,
William White,
Capt. George Lane,
Peter Likekey,
Gilburt Budd,
James Jahcocks,
Gabriel Archer,
Henry Wiltsee,
Petor Drak,

matheuw mcCabe,
Cornelius Tompkins Junior,
Danel Buckbee,
Comfort Chaddick,
Thomas Lewes,
Nathan Lane,
Moses Dusenberry Junior,
Joseph Garrison,
Peter Warren,
Peter Keley,
John Yeouman,
Abraham Croft,
Abraham Marling,
Joseph Bare,
Elisha Budd,
Titus Travis,
Gilbert Oakley,
John Drake,
John Edgar,

Philip Steenburk,
John Knapp,
Isaac Jacocks,
Richard Denny,
Isaac Garrison,
David Henyon,
Isaac Danford,
Thomas Williams,
John Christian,
Jessee Owen,
William Deausenberry,
Solomon Smith,
thomes Brient,
Joshua Tompkins,
Charles Cristian,
Jonathan Miller,
James peney,
Nathaniel Tomkings,
Col. Samuel Drakes."

Cold Spring Village.—This is the largest village in the town or county, and the only incorporated one in it. The Act, incorporating it, was passed by the Legislature, April 22, 1846. Vide chap. 102, Session Laws of 1846. It is twenty miles from Carmel, and one and a half from West Point. It covers a large extent of ground, embracing what is called the Foundry district. The west end of the village, from the store of Lewis Birdsall, to the present steam-boat landing, and some portion of it, north and south of Birdsall's store, on Foundry Street, is made ground, and was formerly a bay; and by filling up, the docks have been extended into the river to their present location. It takes its name from a spring of water which is unusually cold, located on the line of the low and high grounds of the village, at the north-west corner of the door-yard of Henry Haldane, Esq.

It contains 4 churches, 4 clergymen, 2 attornies, 4 physicians, 10 stores, and 4 taverns. If we may be allowed the expression, it is the commercial metropolis of the county, and is the principal freighting depot, on the east side of the river, between the Dutchess and Westchester line. It is the birth place of Lieut· Col. Duncan, of the United States Army, who has rendered his country signal service on the bloody battle fields of Mexico. The old house in which this gallant officer was born, is no longer standing, having been accidentally burned down in 1841. It stood, at the time of his birth, in nearly the centre of Main Street, opposite to the new, large frame building, lately erected by Oliver Elwell, Esq.; the road or street, at that time, running south of Main Street, as it now is. The house, some years since, was moved to the position it occupied when burned down, and was used as a paint shop.

The true name of the Col. is *Duncanson;* and, as his father alleges, when he entered the army, by an oversight, or mistake of the recording clerk in the War Department, at Washington, his name was written *Duncan.* The mistake not being corrected, the Department, as a lawyer would say, "*stuck to the record;*" and the Col., since then, in all his communications to Government, and others, has written his surname, Duncan.

Within the last few years the village has grown rapidly, and is still increasing as fast, perhaps, as any other on the East Bank of the Hudson. The West Point Foundry, located here, has been the main cause of its flourishing condition; and within the last five years its building lots have doubled in value.

Nelsonville.—This village is only a continuation of Cold Spring, and is built on the reverse slope of the hill, on which a part of the former is built. There are a few rods of ground intervening upon the top of the hill, but they will soon be covered with houses. The plot, originally made, embraces both villages. The turnpike leading from Cold Spring to Carmel runs through it. Like Cold Spring, it has greatly increased in population, buildings, and business, within the last few years. It is named after the family of Nelsons, which are numerous in this town, to which is added *ville,* from the Latin "*villa,*" signifying a *village.*

Davenport's Corners.—A small collection of houses, about four miles north of Cold Spring, on the road leading to Fishkill. John Davenport, deceased, kept a store and tavern there, after whom it is named. It is sometimes called "The Old Highland Church," which is located there; but it is more generally called "Davenport's Corners." John Davenport was the father of Elijah J. Davenport, Esq., of Cold Spring.

Griffin's Corners.—A few houses, at the intersection of the Cold Spring turnpike with the old post road, three miles east of Cold Spring. John Griffin resides there, in the old house, built by Gilbert Budd, before the Revolution. It has undergone some repairs, and additions have been made to it, but the old part is still standing.

Break Neck.—A small collection of houses on the east bank of the Hudson, about two miles north of Cold Spring, through which the River-road runs to Fishkill Landing. Many of the best stone-quarries in this town are located here; also the brick-yard of ―— Clark, Esq., and a short distance south of it, the brick-yard belonging to the estate of Daniel Fowler, deceased. These are the only places where brick is manufactured in this town. It takes its name from Break Neck Hill, at the southern base of which it lies; the etymology of the name will be given in the description of the "Hill." Here is a dock, erected by the Harlem High Bridge Company, who rented from Messrs. Howard & Haldane, a few years since, their stone-quarries and the adjacent land, from which a greater part of the stone at this place is shipped.

Dennytown.—A settlement of French people in the central part of the town, about one mile west of Justis Nelson's mill. It was settled by a family of the name of Denny, whose descendants are numerous in this town, and who own a large tract of land in that region. *Town* is derived from the *Saxon* word *tun*, and signifying a *walled or fortified place; a collection of houses inclosed with walls;* in popular usage it means *a township.* It takes its name, therefore, from this family of early settlers.

Hortontown.—A small settlement and district of country in the northern part of the town, about one mile north of the second gate on the turnpike leading to Carmel, and so called from an old and numerous family by the name of Horton, who resided there, many of whose descendants still live there and in the vicinity.

Eel Point.—A small narrow point of land, jutting into the Hudson a few rods above Cold Spring. There is a small bay on the north side of it, a part of which is uncovered at low water, revealing a sandy bottom. In 1810, Henry Haldane, Esq., built the house now standing there, occupied by James Warren, Esq., as a dock and storehouse. The storehouse was taken down in 1845. The Hudson River Railroad passes over the west end of it. It was called "*Eel Point*, near *Sandy Landing.*" Eels congregated there, and its proximity to the sandy shoal just above it, accounts for its name.

Continental Village.—A few houses in the southeast part of the town, one mile from the Westchester line. In the revolution it was the main entrance to the Highlands in this town, and was guarded by a detachment of American troops. Two small forts were erected for its defence, one at the north end of the village on the high ground, the remains of which are still to be seen; and one about a quarter of a mile north-west of it, on the road leading to the Hudson, the foundation of which is also standing. It was burnt by a detachment of British troops in the early part of October, 1777, after forts Clinton and Montgomery had been taken by Sir Henry Clinton. John Meeks was the first settler at this place. The first grist-mill in Philipstown was built at this place by Col. Beverly Robinson, about 1762, and stood a few rods east of where the paper-mill now stands, on the same stream. He also erected a saw-mill and fulling-mill there. Gallows Hill lies to the south of it, in full view, just across the Weschester line, on the east side of which the British troops advanced to the south entrance of the village. In the Revolution, a man by the name of John Strang was caught in the act of enlisting men for the British army, with a commission in his pocket signed by one of the generals. He was tried, condemned to the gallows, and hung upon this hill; hence the name.

Warren's Landing.—The house at this landing was built in 1804 by John Warren, father of the Hon. Cornelius and Sylvenus Warren, of Cold Spring. It is opposite West Point, a little to the west of Constitution Island. About 1798 George Jefford built the dock and a house, which was subsequently torn down.

Mead's Dock.—This place is now called Garrison's Landing. It was built about the year 1814 by Joseph Mead. A store was formerly kept there. It is nearly opposite West Point.

Bross's Landing.—A dock a few rods south of Mead's, built before the last war by a man of the name of Hoyt, a cooper by trade, who built a house, shop, and constructed the dock for the accommodation of freighters in that vicinity and to increase the facilities of his own business.

Hog-back Hill.—A steep eminence at the southwest corner of the flat on which the West Point Foundry is built. It takes its name from its fancied resemblance to the back of that animal.

Vinegar Hill and Mount Rascal.—Two elevated ridges, lying parallel with each other; the former on the east side, and the latter on the west side of the West Point Foundry. They form the sides of the Foundry brook where it empties into the Hudson. It is said they were named by William Youngs, Esquire, who formerly was a *manager* in the Foundry, after two places bearing those names in Ireland.

Bull Hill.—A lofty peak of the Highlands just above Cold Spring, and separated from Breakneck peak by a narrow depression, or slope of land. Although termed a *hill*, it, with Breakneck north of it, is more properly a mountain. But not so thought a *son* of *Erin*, who, being met on the road from Cold Spring to Breakneck by a traveller, was asked, " What mountain is that, my friend ?" To which he replied, " *Sure, sir, an devil a bit* of a mountain is it *a-tal, sir —its Bull Hill.*" According to tradition, a bull had made it his mountain home, from which, at night, he would descend to the low grounds in its vicinity, and commit sundry depredations in corn fields, meadows, and grain fields. The neighbors formed an "*alliance,*" offensive and defensive, against this bold and ruthless mountain robber, determined to pursue him to his strongholds, and effect his capture or destruction. They chased him from this *Hill* to the one immediately north of it, where, being hard pressed by dogs and armed men, he was " impelled and propelled down a precipice and through the *chaparal,*" by which *Sam Patch*-like leap, he broke his neck ; whereupon his pursuers immediately christened the hill from which they started him " *Bull Hill,*" and the one where they captured this midnight *guerilla chief,* " Breakneck." The word *hill* is from the *Saxon hyl,* and means " a natural elevation of land, or a mass of earth rising above the common level of the surrounding land."

Break Neck.—Another lofty peak of the Highlands, just north of Bull Hill, the name of which has been explained in the description of that locality, above. A tunnel is now being cut through the western flank of this peak, for the Hudson River Railroad. The western flank protrudes itself almost into the river, through the centre of which runs the dividing line of Dutchess and Putnam Counties. On the south side of this peak, and within Putnam, within a few feet of its apex, " *St. Anthony's Face,*" so celebrated in the history of the Hudson's scenery, once peered out and over the rocky battlements below, gazing, as it were, at the eternal *ebb* and *flood* of the mighty current that breaks with unceasing fury against the lofty parapet that supported it. Thousands of the travelling community on board of steamboats, with glass in hand, have turned their eyes on passing *Break Neck*, to gaze upon the stern and rugged features of *St. Anthony's Face ;* but the venerable patriarch, destined like everything of earth, has passed away, and is numbered among the things that were.

In the summer of 1846, Capt. Deering Ayers, who was engaged in the service of the Harlem High Bridge Company, by one fell blast, detached an immense block of granite weighing nearly two thousand tons, and shivered in atoms the majestic brow and weather-beaten features, of the venerable *mountain hermit.* Nero, the last Roman emperor of the family of the Cæsars, set fire to Rome, merely, as it is reported, that he might have a real representation of the conflagration of Troy, and *fiddled* while it was burning. We are not informed whether *Ayers,* like Caius Marius amid the ruins of Carthage, smiled over the wreck that lay shattered around him, or evinced sorrow at his wanton demolition of nature's sculpture ; but the act was *vandalic,* in the extreme, to the true lover of Nature's works ; and the more so, as the stone was utterly useless to those who sought it in its mountain-home.

"From his eyrie, that beaconed the darkness of heaven,"

and for ages, *he* had looked abroad upon the restless and ever-agitated world, defying the warring elements of nature and the tooth of time—majestic in the solitude of his mountain-home, he stood an admired specimen of nature's mechanism,

> "A man without a model, and without a shadow."

> "O, woe to Mammon's desolating reign,
> We ne'er shall see his like on earth again."

The tragedy was not ended with the destruction of *St. Anthony's Face*, for with the same terrible and destructive agent with which the venerable saint was hurled from his airy pedestal, the poor unfortunate Ayers met with an untimely death. Some months after the scene at Break Neck, Ayers was engaged on Staten Island, blasting rocks. Having set fire to the fusee, he retired, but the blast not going off within the usual time, he returned to it, and having commenced working with it, it exploded, blowing the *celebrated blaster* of St. Anthony's Face into an hundred fragments.

The greedy, sordid, and avaricious spirit of man is making sad havoc among the beautiful mountain scenery of the Hudson. Where it will stop, is more than we can tell. The sound of the ax and the railroad excavator's pick, with the steady click of the quarryman's hammer, is daily sounding in our ears, and slowly, though steadily, performing the work of demolition.

> "Rock and tree, and flowing water,"

are alike the subjects of this railroad tariff—a sort of *steam custom-house* tax; and with the stereotyped plea of *utility* in one hand, the utilitarian of the nineteenth century, with the other, grasps without remorse the beauty of the richest landscape, and all that is noble and sublime in the scenery of the natural world.

This *chiseling* of *the Great Architect*, bearing such a striking resemblance to the human face, was named in honor of "*St. Anthony the Great;* first institutor of monastic life; born A.D. 251, at Coma, in Heraclea, a town in Upper Egypt." At the base of this mountain peak, on the eastern shore of the Hudson, begrimed with smoke, dust, and powder, the mutilated features of the celebrated *Face of St. Anthony* now lie,

"And none so poor to do them reverence."

"Sic transit gloria mundi."

Cat Hill.—A large rocky hill, about two miles east of Cold Spring, near Justus Nelson's mill. In the earliest settlement of this town, it was the resort of *wild cats.* As the country became more thickly settled and cleared up, those animals were entirely exterminated; and dropping the word *wild,* the people named it as above. Its wild, rugged aspect, would have justified the inhabitants in retaining the first word, as the name would not have belied its appearance.

Sugar Loaf Mountain.—A lofty peak of the Highlands about two miles southeast of West Point, near the "*Robinson* House." Its altitude, as taken by Lieut. Arden, late of the U. S. Army, is 800 feet above the level of the Hudson. Its shape is conical, resembling a loaf of sugar, and hence its name.

Anthony's Nose Mountain.—This is the highest peak of the Highlands, on the east side of the Hudson, in this town. It is situated at the entrance of the

Highlands, and is about 1100 feet in height. It is in the south-west part of the town, near the line of Westchester and Putnam, and opposite Fort Montgomery, on the west bank of the Hudson. From the base of this peak, a large boom and chain extended, in 1776, to Fort Montgomery. This was the second obstruction attempted in the Hudson, to prevent the British from ascending it. The first was at Fort Washington, below the Highlands in Westchester County; the third at West Point and Constitution Island; and the fourth at Pallopel's Island, at the south entrance of Newburg Bay, extending to Plum Point, on the west bank of the river.

From the Journal of the Committee of Safety, we extract the following, respecting the chain at *Anthony's Nose* and Fort Montgomery:

Nov. 30th., 1776. In perfecting the obstruction between St. Anthony's Nose on the eastern shore and Fort Montgomery, we endeavored to avail ourselves of the model of that which had proved effectual in the river Delaware, and were assisted by the advice and experience of Capt. Hazelwood, but the great length of the chain (being upwards of 1800 feet), the bulk of the logs which were necessary to support it, the immense weight of water which it accumulated, and the rapidity of the tide, have baffled all our efforts; it separated twice after holding a few hours.

"Mr. Machen, the Engineer at Fort Montgomery, is of opinion, that with proper alterations it may still be of service in another part of the river, and we have, with Gen. Heath's concurrence, directed him to make the trial.—But we have too much reason to despair of its ever fully answering the important purpose for which it was intended. A like disappointment we are informed happened at Portsmouth. Gen. Heath, on a conference with Gen. Clinton, has been pleased to recommend the obstruction of the navigation in this part of the river by cassoons," &c.

Gordon, in his Gazetteer, states, that the cost of this

chain at £50,000, continental money; that it was made of iron 2 or 2 1-2 inches thick, was 1,800 feet in length, and weighed 50 tons.

From a fancied resemblance of this peak to the human *nose*, and in honor of St. Anthony, it received the above appellation before the Revolution. There were two redoubts on this mountain, a short distance apart. They were intended to guard the entrance of the Hudson into the Highlands, and as an additional security to the *chain*. It is said that the manufacture of this chain, with the cost of placing it across the river, exhausted the Continental treasury; and so far as any good was effected by it, Congress might about as well have caused a roll of twine to be stretched in its stead. Its own weight parted it twice, and when the leading English ship struck it, it broke with the facility of a pipe-stem.

Pine Hill.—An eminence near the Mansion House of Mrs. Mary Gouverneur, so called from pine timber growing upon it. Philips' Quarries are located there.

Stony Point.—A rocky peninsula, nearly half a mile north-west of the village of Cold Spring, and stretching into the Hudson about one fourth of a mile. The west end is an entire mass of granite rock, and has been quarried successfully by Alex. Anderson & Co., to whom it now belongs.

It is steep at the west end, with a sufficient depth of water to float vessels of the largest class navigating the Hudson.

Whiskey Hill.—A small eminence on the old road leading from Continental village to Garrison's Landing. During the Revolutionary war, some soldiers were carting a hogshead of whiskey from Continental village to West-Point, for the use of the garrison at that post. On reaching nearly the top of this hill, the blocks in the hind part of the cart slid from their positions, and the hogshead, smashing the tail-board into pieces, rolled to the foot of the hill, where coming in contact with a large stone, it bursted to the deep chagrin of the soldiers, who were anticipating a hearty dram of it on their arrival at the post. They immediately christened it " Whiskey Hill," which name it has ever since retained.

Fort Hill.—A lofty eminence, the timber of which has been cut off, a few hundred rods east of the mansion-house of Judge Garrison. Two redoubts were erected on it in the Revolution, one at the North end called " North Redoubt," and the other at the south end called " South Redoubt." This redoubt may have been the one spoken of in Gen. Heath's order of Dec. 3d, 1780, as the " Middle Redoubt." There were works thrown up on Sugar Loaf Mountain ; and by way of distinguishing the two, the one on the South end of this hill may have been called the " Middle Redoubt."

Extract from " Revolutionary Orders."
" GEN. HEATH'S ORDERS.
"Head Quarters, West Point,
Decem. 3d, 1780.

" Brig. Gen'l Huntingdon will please to assign one Regiment of the Conn. Line to the defence of the North Redout, one to the Middle Redout, and one to the works on Constitution Island, which works are to be considered as the posts of those three Regiments in case of alarm ;- the other Regiments of the Line, in such case, are to be held in readiness to act as circumstances may require.

" The 4th Mass'ts Brigade is assigned to the defence of Fort

Clinton and its dependencies ; the 2nd Brigade to the defence of Forts Putnam, Willis, and Webb ; Col. Shepard's and Col. Bigelow's to the former, Col. Vose's to Fort Willis, and Lt. Col. Commandant Smith's to Fort Webb : the 1st and 3d to act as circumstances may require, and, on all alarms, to form on their Brigade Parades, ready to receive orders.

" The Connecticut Line is to mount a Captain's Guard at the Continental Village for the security of the public stores, and guarding that avenue into the Highlands."*

The Sunk Lot.—This name is given to a tract of land in the east part of the town, containing about 1300 acres, belonging to Joel Hamilton. Its northern termination is near the Cold Spring turnpike, about one and a half miles south west of Griffin's Gate, and extends south nearly to the former residence of Joel Bunnell, Esq. Its location is low, apparently *sunk* down ; and hence the name.

Constitution Island.—This island, projecting half way across the Hudson, forms its elbow nearly opposite West Point. Its western side is formed by steep and inaccessible precipices ; on the east, between it and the main land, is a large marshy, flaggy meadow, which, within a few years past, has been partly drained by ditches cut through it. This island is, probably, about two miles in circumference, and half a mile wide from north to south. It is covered with timber of an inferior kind, and uncultivated except on the southern and eastern edges. The entire marsh meadow contains about 300 acres, and the island about 250.

This island, previous to, and at the commencement

* This extract is given to show the Military localities of West Point at that early day.

of, the Revolution, was called "*Martelaer's Rock Island*;" but after a fort was erected here in 1775, it was more often called Constitution island, by which name it is now known. The fort was called "Fort Constitution." In the correspondence between the officers of the army and the New York Committee of Safety, and also with the Continental and Provincial Congress, it is sometimes written "*Martles' Rock*," and also "*Martyr's* Beach."

From the most accurate information that we have been able to obtain, this island was called after a Frenchman by the name of *Martelair*, and who, probably, resided on it with his family. A family, bearing that name, were early settlers at Murderer's Creek, in the town of New Windsor in Orange co.; and were murdered by the Indians about the year 1720. It may have been the same family who previously resided on this island, or a branch of it. The Provincial, as well as Continental Congress, early saw the necessity of fortifying the Hudson river to prevent its ascent by the enemy, and thus keep open the communication between the eastern and middle States. The Continental Congress moved first in the matter, but the published record of its proceedings does not disclose the date.

On the 18th of August, 1775, the Provincial Congress of New York passed the following resolution:

"Resolved and Ordered, That the Fortifications formerly ordered by the Continental Congress, and reported by a Committee of this Congress, as proper to be built on the banks of Hudson's River, in the Highlands, be immediately erected. Mr. Walton dissents. And that Mr. Isaac Sears, Mr. John Berrien, Col. Edward Flemming, Mr. Anthony Rutgers, and Mr. Christopher Miller, be Commissioners to manage the erecting and finishing the Fortifications. That any three or more of them be empowered to act, manage, and direct the building and finishing thereof."

The "Fortifications in the Highlands" embraced, not only those to be erected on Constitution Island, but also those afterwards erected on the north and south sides of Poplopen's Kill, called Forts Montgomery and Clinton. These were the main works, while redoubts were built on the neighboring eminences, on the east side of the Hudson; two on Redoubt Hill, called North and South Redoubt, just east of Judge Garrison's residence, two on Sugar Loaf Mountain; and one on Anthony's Nose Mountain. Col. Edward Flemming and Capt. Anthony Rutgers, notified the Provincial Congress that they could not attend to the duties of Commissioners; on the 22nd of August, in the same year, Capt. Samuel Bayard and Capt. William Bedlow, were appointed in their stead.

The Provincial Congress employed Bernard Romans, who held a commission as Engineer in the British Army, to construct the "Fortifications in the Highlands." By order of the Committee of Safety, he commenced operations on the 29th of August, on Constitution Island; and on the 12th October, 1775, he applied to the former body for a commission, with the rank and pay of Colonel.

"Fort Constitution, October 12, 1775.
"Honourable Gentlemen:
"By order from the Committee of Safety, I am up here for the purpose of constructing this fort; said gentlemen gave me their words that I should be appointed principal En-

gineer for this Province, with the rank and pay of Colonel. As I have been now actually engaged in this work since the 29th of August last, I should be glad to know the certainty of my appointment, and therefore humbly pray that my commission may be made out and sent. I have left the pursuit of my own business, which was very considerable, and endangered my pension from the Crown, by engaging in our great and common cause. These matters considered, I hope my request will be thought reasonable, and therefore complied with. I remain, with the utmost respect, honourable Gentlemen, your most obedient humble servant,

"B. Romans."

Romans and the Commissioners soon became involved in an unpleasant dispute about the construction of the works on this island. Romans claimed the right, by virtue of his office, to build the works according to his own furnished plan; and pointedly told the Commissioners that they had no right to interfere with his operations; that their business was to furnish him with men and money, reserving their condemnation or approval until the Fortifications were finished.

The Commissioners, on the other hand, claimed the right, as superintendents, to approve or reject his plans, and direct the mode of operations, contending, that his duty was to work according to their directions. They objected to his plans, as involving too much expense to the State. A long epistolary correspondence followed, with drafts, reports, and estimates, for which we have not room to gratify the reader, as our work is limited to a given number of pages. A condensed view of the "Reports" and "Plans" of the Fortifications to be erected on this Island, made by Romans to the Commissioners, and through them to the two Congresses, is as follows:

On the south side of the Island, he proposed to erect five block houses; barracks, 80 by 20 feet; store-houses and guard-room, 60 by 20 feet; "2,400 perches of stone wall, each perch containing 16½ feet in length, 18 inches high, by 12 wide;" five batteries, mounting 61 guns and 20 swivels; a fort, with bastions, and a curtain, 200 feet in length; a magazine; and estimated the entire costs, materials of every description, with "labour of, and provisions for, 150 men for four months, 26 days to the month, at an average of 3s. per day, at £,4,645 4s. 4d." The cost of ordnance was not included in the above sum.

In the Revolution, this Island belonged to Mrs. Ogilvie and her children. She was the widow of Capt. Ogilvie, a British officer in the French and Indian war; and grandmother, as we have been informed, of Mrs. Mary Gouverneur. The Committee of Safety, when about to fortify it, applied to Beverly Robinson, offering to purchase it. The following is the correspondence between them respecting the purchase of sufficient ground to erect the works on:

"In Committee of Safety.
"New York, Sept. 19th, 1775.

"Sir—By order of the Continental Congress, founded on the necessities of the present times, the Provincial Congress of this Colony has undertaken to erect a fortification on your land, opposite to the West Point, in the Highlands. As the Provincial Congress by no means intend to invade private property, this Committee, in their recess, have thought proper to request you to put a reasonable price upon the whole point of dry land; or island, called Martelair's Rock island; which price, if they approve of it, they are ready to pay you for it.

"We are, sir, your humble servants.
"To Beverly Robinson, Esqr., at his seat in the Highlands."

"In Provincial Congress, New York, 6th October, 1775.

"A letter from Beverly Robinson, Esq., was read and filed, and is in the following words, to wit:

"Highlands, October 2nd, 1775.

"Sir—Your letter of the nineteenth of September, I received a few days ago, in answer to which, I must inform you that the point of land on which the fort is erecting, does not belong to me, but is the property of Mrs. Ogilvie and her children. Was it mine, the public should be extremely welcome to it. The building a fort there can be no disadvantage to the small quantity of arable land on the island. I have only a proportion of the meadow land, that lays on the east side of the island.

"I am, sir, your most humble servant,
"BEV. ROBINSON.

"To John Haring, Esqr., Chairman of the Committee of Safety, at New York."

The Hudson River Railroad crosses the east end of this island, and on the south-east part of it, is cut through a gravel hill, on which was erected a *bumbert* battery in the Revolution.

Opposite to West Point, embowered among trees and shrubbery, and surrounded by the eminences on which the "fortifications" were built, stands the sequestered and rural country seat of Henry W. Warner, Esq., Counsellor, &c., called "Wood Crag." The kitchen part of this mansion is a fragment of the old barrack erected in 1775. The remains of Fort Constitution at the water's edge, are still to be seen. This island, with the marsh meadow east of it, belongs to H. W. Warner, Esq.

Ardenia.—This beautiful country seat is the residence of Richard D. Arden, Esq., and is situated on the east bank of the Hudson, about one fourth of a mile north of the "Robinson House."

Highland Grange is the name of the mansion house built by Capt. Frederick Philips, deceased. It is a charming spot, located on the immediate bank of the Hudson, opposite to West Point. About half a mile north of Highland Grange, is the summer residence of Henry De Rhams. It is a lovely spot, with an extensive view north, south, and west. We do not know that it has been christened by any rural name.

Under Cliff.—This is the romantic and beautiful country seat of the great lyric poet of our country, Gen. George P. Morris. There is a romantic truth in the name, for it is well-nigh under one of the bold, rugged, and frowning cliffs of Bull Hill, at its southwest side. Here, the lover of the grand and sublime—the amateur of the mystic science of nature's works—may scan, on a most stupendous scale, those immoveable bulwarks, against which " the artillery of a thousand armies might roar out their ineffectual vengeance," while " the parapet would laugh in scorn at the power of battle."

From no residence in the Highlands, perhaps, can such an extended view be had, as at Under Cliff. " To the right, to the left, in every direction, tower the rocky pinnacles of the Highlands, whose giant forms seem separated by the hand of Omnipotence, to make way for the quiet Hudson, as she hastens to pay her tribute to her monarch, the ocean." To the right and north, Bull Hill and Break Neck stand, like weather-beaten sentinels, to guard the further encroachment of the mighty current on the east, as it surges from the broad and ample bay above, through the Highland pass. To the west looms up Butter Hill and Crow's Nest, casting their sombre shadows far into the Hudson; while to the south, Fort Putnam, at an elevation of 500 feet above the river, with its massive walls, still venerable in its ruin, stands "to give an ocular demonstration of the untiring industry and hardy enterprise of the heroes of '76." The scenery of the Hudson, in this vicinity, is unequalled, and "bears nature's grandest imprint." The Rhine, in Germany, is said to resemble it more than any other, but does not equal it. This mansion is a few hundred rods north of Cold Spring. Here, beside this mighty river in its hour of glory, at sunset—tinging with Eden dies the most gorgeous scenery the eye ever rested upon—" where the rock throws back the billow, brighter than snow"—is the spot most fitting for the wrapt mountain-bard to tune his lyre, and chant an anthem to the sylvan deity of the place.

Round Pond.—A small body of water located on the land of Daniel Smith, Esq., in the south part of the town; and covers about three acres, containing perch and trout. It is circular or round, and hence its name.

Cat Pond.—A small sheet of water at the base of Cat Hill, on the land of Mrs. Mary Gouverneur, covering about two acres, on the west side of the "Old Post-Road," near Justus Nelson's mill. It takes its name from its contiguity to Cat Hill.

The Robinson House.—This mansion, around which the stirring incidents of the revolution have flung such an interesting and melancholy interest, is situated in the south-west corner of this town, upon the water lot formerly owned by Col. Beverly Robinson, about 400 yards from the Hudson, in a straight line, and at the base of Sugar Loaf mountain. It is about two miles south-east of West Point, and four miles south of the village of Cold Spring. We gazed long and intensely at this memorable building ere we entered within its walls, on our first beholding it. Its grounds and halls have been hallowed by the tread and presence of the "*Father of his Country,*" by Knox, Greene, Putnam, Steuben, Kosciusko, Heath, Parson, McDougal, and many others, in "times that tried men's souls." And even while the patriot of his own country, and the lover of liberty from another—Lafayette—rested beneath its roof, it has also held the dark, clutching, sordid traitor, Benedict Arnold. It was here, in the upper back room of the main building, where Arnold began and completed those sketches and drawings of the fortifications and works at West Point, which subsequently cost the youthful and accomplished Andre his life. Here he perfected and finished the requisite evidence of his allegiance to the British King, blackening the page of our country's history by a perfidy that is without a parallel, and unlike Judas, he refused to weep, but singing the song of his own infamy, he sank " like mad Ophelia on the wave, singing as he sank."

The confidence that Washington reposed in his confidence was unbounded, or he would hardly have entrusted him with so important a post as West Point, which was the key that would unlock the southern door of the northern department. And how could it be otherwise? Who could doubt the fidelity and patriotism of the leader of that Spartan band of heroes who marched in the depth of a rigorous winter from the cold, bleak, and barren frontiers of Maine to the rock-bound citadel of Quebec? Who could entertain a suspicion of the man, who, after marching the foremost of his little band two hundred miles through the snow-clad forests of Maine, over rocks and precipices, and the inhospitable deserts of Lower Canada, where the foot of the white man had never passed, had sat down to satisfy the cravings of hunger on the body of a dead camp dog on the banks of the Chaudiere? He who headed the forlorn hope at the storming of Quebec, where his leg was shattered by a musket ball; who had poured out his life-blood like water on the plains of Saratoga and Stillwater for his country; who had fronted the cannon's mouth, charging up to their very muzzles amid the storm of iron hail that so dreadfully wasted his followers; could such a man be trusted? Washington felt that it was wickedness to doubt, and he gave to Arnold the command of West Point and its out-posts.

But when the dreadful truth was disclosed, it wrung his great spirit with an anguish that his officers had never before witnessed: and the question he asked of Lafayette, "*Who can we trust now*," shows the extent of confidence reposed by Washington in the patriotism of that fallen hero of our early struggle—Benedict Arnold.

The following acrostic is about as severe and sarcastic as it is possible to express in the English language:

"Born for a curse to virtue and mankind,
Earth's broadest realms can't show so black a mind;
Night's sable veil your crimes it cannot hide,
Each are so great, they glut the historic page;
Defam'd your memory shall for ever live,
In all the glare that infamy can give;

Curses of all ages shall attend your name,
Traitors alone shall glory in your shame.

" Almighty vengeance sternly waits to roll
Rivers of brimstone on your treacherous soul:
Nature looks back, with conscious sorrow sad,
On such a tarnish'd blot as she has made—
Let hell receive you riveted with chains,
Doom'd to the hottest focus of its flames."

Three buildings joined to one another compose the mansion. Nearest to the river is the farm-house, one story high, and connected to it, on the east, are the two main buildings, two stories high, with a piazza extending along the north, east, and southerly sides of the building nearest the Sugar Loaf Peak, and on the south side of the centre building. The house and lands attached now belong to Richard D. Arden, Esq., and are occupied by his son, Lieut. Thomas Arden, late of the U. S. Army, and an officer in the Florida war.

" The same low ceiling, large and uncovered joists, the same polished tiles around the fire-places, and the absence of all ornament which marks the progress of modern architecture, preserve complete the interest which the stirring incidents of that period have flung around the Robinson House."

In the centre building is the large dining-room, where the traitor, with his wife, and two of Washington's aids-de-camp, were at breakfast, when a messenger dashed up to the door and handed him a letter, which the stupid Jamieson had forwarded by express to Arnold, informing him of the arrest, and discovery of the papers. We have stood within this room, we have planted our feet upon the broad staircase that the avaricious traitor mounted "in hot haste," after reading Jamieson's letter, as he flew to the chamber of his wife in the second story of the eastern main building fronting to the north, where he "disclosed to her his dreadful position," urging her to burn all his papers, and informed her "that they must part for ever."

This house has been kept from dilapidation and decay by repairs, when needed, but in no way has it been changed from its original appearance. It has been roofed from time to time, as often as the wear and tear of the elements have rendered it necessary; a new piazza has been added in the place of the former one, but no alterations have been permitted, either inside or out, that has changed, in the least, its original shape and appearance.

Beverly Robinson, who built it about 1750, was a Major in the British Army, under the gallant Gen. Wolfe, in the battle upon the Plains of Abraham. The lands, originally attached to the mansion, are of an excellent quality, and numbered one thousand acres, under a good state of cultivation. He married a beautiful, amiable, and accomplished lady, a descendant of the original patentee of Putnam County, by whom he acquired large tracts of land, and then retired from the army to the enjoyment of that domestic happiness upon his estates, which a rural life and such a partner are so well calculated to secure. He was for several years Supervisor of this town, and took an active part in everything that concerned its interests. The lands he acquired by his marriage with Miss Philips, was the water-lot, four miles square, and on which the "Robinson House" stands; the first long-lot adjoining the water-lots on the east; and the short-lot in the north-east part of the county.

The reader, by turning to the diagram in the previous part of this work, will see the form of the partition made to the three heirs to whom it was devised, and those lots or divisions which Col. Robinson acquired with his wife.

While residing at "*Beverly*," in this quiet and secluded retreat, where nought is heard but the sighing of the breeze, the murmurs of the rolling Hudson, the song of the robin, and the whoop of the whippo-will —surrounded by every comfort that the heart can desire, and dispensing a generous hospitality—the storm that had been long gathering between the Mother-Country and her Colonies burst forth, and he was summoned to the field, by virtue of the right of England's King, to demand the services of his native-born subjects in time of need. He obeyed the call with great reluctance, and it is said, pled hard to be allowed to remain in the bosom of his family, and in the quiet enjoyment of his rural pursuits. But the constant and unceasing solicitations of his influential and English friends, reminding him of his oath of allegiance to his King and country—that he was a native-born subject, and when his country had called him in days gone by, his soldier-like heart faltered not on the Heights of Abraham, where he became enamored of her glory—at last prevailed under a sense of that stern system of teaching, which impresses on the soldier, as the first duty, obedience to superior authority.

Having removed his family to New York, he departed from his long loved "*Vallambrosa*," to the

British army, in which he received a Brigadier-General's commission. His family never returned to "Beverly," where they had spent so many years of unalloyed happiness; but when the British army under Gen. Clinton moved up the Hudson after the battle of Fort Montgomery, it is said, he visited, and for the last time, his house, to which he was destined never more to return.

Such men at this day are sometimes called "*Tories.*" But is the charge just or true? With a solemn regard for truth, we think not. When the reader shall have patiently examined all the facts up to the commencement of the Revolution in the life of Beverly Robinson, we think he will concur with us, that he was not a *tory*, nor can he be brought within the meaning of the word, as understood by the native-born patriots of the Revolution. Webster says that the word "*tory*" is "said to be an Irish word, denoting a robber." That "in America, during the Revolution, those who opposed the war, and favored the claims of Great Britain, were called *tories.*" But this latter meaning and understanding of the word, as given by Webster, was, and could only be applied to native-born citizens of this country, and not to those born and brought up in the Mother-Country, who had passed more than half their lives in her standing armies. The latter class of persons, at the breaking out of the war, were just as much Englishmen, and subject to the laws and government of that country, although they had resided a few years in the colonies, as if they had always remained in England. How stands the case with respect to Col. Robinson? He was not a native-born citizen of this country, and at the time he returned to the British army, we had not a national existence. To whom then did he owe alliegance? not to us, for we did not exist at that time as an independent nation; not to the country, for he was not born here; not to the land, for that he only held in right of his wife, in whom the title was vested; and when it was confiscated by an act of the legislature, the reversionary interest was not affected, for "in 1809, John Jacob Astor bought the reversionary interest of the lands acquired by Beverly Robinson by his marriage with Miss Philipse, and also those of Major Roger Morris, who married Miss Mary Philipse, sister to Col. Robinson's wife, off the heirs of both for $100,000." For this Mr. Astor received from the State, 10 years after, the *small sum* of $500,000! Besides, after the battle of Quebec, in 1759, and the Treaty of Peace in 1763, whereby England became possessed of Canada, he had only retired, temporarily, as an officer, from the army, but was liable to be called on at any time by the British government, in case of war, to resume his rank therein. He had been educated to the profession of arms, and " two generations of the Robinson family " had held commissions in the service of their country, and bore arms in wars waged by English kings.

All offices, civil and military, that he had occupied, he had held under the government and laws of Great Britain.

Had he been found at the head of an American regiment at the commencement of the war, by its rules and regulations, had he been taken, his government would have caused him to be shot. But the case is very different when applied to a native-born citizen.

If an officer of his rank had left our army and resided a few years in Mexico, and during the present war with that nation should be captured in time of battle fighting at the head of his regiment, his chance for life would be feeble before a court martial ordered by Scott or " Old Rough and Ready." The law of nations does not permit a man to change his allegiance while his country is at war with another. The law makes it his duty in such an event to return, if it be possible, and offer his assistance in the hour of her need. It matters not whether the war be just or unjust on the part of his country, his inability to change his allegiance still continues. "It is the doctrine of the English law, that natural-born subjects owe an allegiance which is intrinsic and perpetual, and which cannot be divested by any act of their own." The fact also, that some native-born citizens of Great Britain, who had been officers in her armies, assisted in achieving the liberties of our country, alters not the case. Had they fought for their own country instead of ours, they would not, by so doing, have been tories. In no view of the case, therefore, can we regard Beverly Robinson as one.

The following extract is from the pen of the late Dr. Timothy Dwight, who, in 1788, was Chaplain of the army, and stationed at West Point, but resided with Gen. Putnam, who held his head-quarters at the " Robinson House :"

"A part of this time I resided at the head-quarters of General Putnam, then commanding at this post; and afterward of General Parsons, who succeeded him in the command. These gentlemen lodged in the house of Col. Beverly Robinson; a respectable native of Scotland, who married a lady of the Phillips family, one of the wealthiest and most respectable of the province of New York. With this lady Col. Robinson acquired a large landed estate lying in Philipstown, Fredericktown, and Franklin, as they are now called; and for the more convenient management of it planted himself in this spot. Here he had a spacious and convenient mansion, surrounded by valuable gardens, fields, and orchards, yielding everything which will grow in this climate. The rents of his estate were sufficient to make life as agreeable as from this source it can be. Mrs. Robinson was a fine woman; and their children promised every thing which can be expected from a very hopeful family. His immediate friends were, at the same time, persons of the first consequence in the province.

"When the revolutionary war broke out, Col. Robinson was induced, contrary as I have been informed to his own judgment and inclination, by the importunity of some of his connections, to take the British side of the question. To him it appeared wiser and safer to act a neutral part, and remain quietly on his estate. The pressure, however, from various sources was so strong against him, that he finally yielded, and carried his family with him to New York, and thence to Great Britain. His property was confiscated by the legislature of New York, and his family banished from their native country. It was impossible for any person, who finds an interest in the affairs of his fellow-men, and particularly while residing in the very mansion where they had so lately enjoyed all which this world can give, not to feel deeply the misfortunes of this family. Few events in human life strike the mind more painfully than banishment; a calamity sufficiently disastrous in the most ordinary circumstances, but peculiarly affecting when the banished are brought before us in the narrow circle of a family; a circle, the whole of which the eye can see, and whose sufferings the heart can perfectly realize. Peculiarly is this true, when the family in question is enlightened, polished, amply possessed of enjoyments, tasting them with moderation, and sharing them cheerfully with their friends and neighbors, the stranger and the poor."

The circumstances attending the flight of Arnold from this house, and the arrival of Washington soon after, is described by one who visited it in 1840, and is extracted from the *Knickerbocker* for Sept. of that year:

"The Commander-in-chief, at the time of the capture, was on his way from Hartford, and changing the route which he had first proposed, came by the way of West Point. At Fishkill he met the French minister, M. de la Luzerne, who had been to visit Count Rochambeau at Newport, and he remained that night with the minister. Very early next morning he sent off his luggage, with orders to the men to go with it as quickly as possible to 'Beverly,' and give Mrs. Arnold notice that he would be there at breakfast. When the General and his suite arrived opposite West Point, he was observed to turn his horse into a narrow road that led to the river. Lafayette remarked, 'General, you are going in a wrong direction; you know Mrs. Arnold is waiting breakfast for us.' Washington good-naturedly remarked: 'Ah, I know you young men are all in love with Mrs. Arnold, and wish to get where she is as soon as possible. You may go and take your breakfast with her, and tell her not to wait for me: I must ride down and examine the redoubts on this side of the river.' The officers, however, with the exception of two of the aids, remained. When the aids arrived at 'Beverly,' they found the family waiting; and having communicated the message of General Washington, Arnold, with his family and the two aids, sat down to breakfast. Before they had finished, a messenger arrived in great haste, and handed General Arnold a letter, which he read with deep and evident emotion.

"The self-control of the soldier enabled Arnold to suppress the agony he endured after reading this letter. He arose hastily from the table; told the aids that his immediate presence was required at West Point; and desired them so to inform General Washington, when he arrived. Having first ordered a horse to be ready, he hastened to Mrs. Arnold's chamber, and there, with a bursting heart disclosed to her his dreadful position, and that they must part, perhaps for ever.* Struck with horror at the

---

* We also visited this chamber, which remains unaltered. Over the mantel is carved in wood-work: "G. Wallis, Lieut. VI. Mass. Regt."

painful intelligence, this fond and devoted wife swooned, and fell senseless at his feet. In this state he left her, hurried down stairs, and mounting his horse, rode with all possible speed to the river. In doing so, Arnold did not keep the main road, but passed down the mountain, pursuing a by-path through the woods, which Lieutenant Arden pointed out, and which is now called '*Arnold's Path.*' Near the foot of the mountan, where the path approaches the main road, a weeping willow, planted there no doubt by some patriot hand, stands, in marked contrast with the forest trees which encircle and surround it, to point out to the inquiring tourist the very pathway of the traitor.

"In our interesting visit, we were accompanied by the Superintendent, Major Delafield, and in the barges kindly ordered for our accommodation, we were rowed to 'Beverly Dock,' and landed at the spot where Arnold took boat to aid his escape. He was rowed to the 'Vulture,' and using a white handkerchief, created the impression that it was a flag-boat: it was therefore suffered to pass. He made himself known to Captain Sutherland, of the Vulture, and then calling on board the leader of the boatmen who had rowed him off, informed him that he and his crew were all prisoners of war. This disgraceful and most unmanly appendix to his treason, was considered so contemptible by the Captain, that he permitted the man to go on shore, on his parole of honor, to procure clothes for himself and comrades. This he did and returned the same day. When they arrived in New York, Sir Henry Clinton, holding in just contempt such a wanton act of meanness, set them all at liberty.

"When General Washington reached Beverly, and was informed that Arnold had departed for West Point, he crossed directly over, expecting to find him. Surprised to learn that he had not been there, after examining the works he returned. General Hamilton had remained at Beverly, and as Washington and his suite were walking up the mountain road, from 'Beverly Dock,' they met General Hamilton, with anxious face and hurried step, coming towards them. A brief and suppressed conversation took place between Washington and himself, and they passed on rapidly to the house, where the papers that Washington's change of route had prevented his receiving, had been delivered that morning; and being represented to Hamilton as of great and pressing importance, were by him opened, and the dreadful secret disclosed. Instant measures were adopted to intercept Arnold, and prevent his escape, but in vain. General Washington then communicated the facts to Lafayette and Knox, and said to the former, more in sorrow than in anger, 'Whom can we trust now?' He also went up to see Mrs. Arnold; but even Washington could carry to her no consolation. Her grief was almost frenzied; and in its wildest moods, she spoke of General Washington as the murderer of her child. It seemed that she had not the remotest idea of her husband's treason; and she had even schooled her heart to feel more for the cause of America, from her regard for those who professed to love it. Her husband's glory was her dream of bliss—the requiem chant for her infant's repose; and she was found, alas! as many a confiding heart has oft been found,

"'To cling like ivy round a worthless thing.'"

The following extract, concerning the traitor's career subsequent to his treason, is taken from a writer to the New Haven *Palladium*. The great moral lesson which his life affords, cannot be impressed too deeply upon the minds of the young and rising generation of our country:

"The close of Arnold's ignominious career was characterized by the loss of caste and the respect of everybody. A succession of personal insults and pecuniary misfortunes followed his treason, and deep abiding retribution was fully meted out to the degraded culprit long before he died.

"An elderly lady of cultivated mind resides in Massachusetts, whose early social intimacy with Arnold and his family at St. John's, New Brunswick, gave her peculiar opportunities for knowing many details concerning the close of his miserable career. Subsequently to the termination of the Revolutionary war, and after the perpetration of various atrocities against his countrymen, Arnold went to England and received a commission in the British army. He was frowned upon by the officers, and everywhere received with contempt, if not indignation. Various public insults were offered to him, and in private life he was the object of perpetual scorn.

"Soon after Arnold threw up his commission in disgust, and removed to St. John's. He there engaged in the West India trade, becoming as notorious for his depravity in business as he had been false to his country; his integrity was suspected at various times, and on one occasion, during his sudden absence, his store was consumed, upon which an enormous insurance had been effected. The company suspected foul play, and a legal contest was the result. During this painful scene his family were greatly distressed, and the lady to whom allusion has been made, and who resided near Arnold's house, was requested to go and pass that trying season with them. That request, in the fair hand-writing of Mrs. Arnold, until recently, was in my possession, as well as a copy of a satirical handbill, describing Arnold's life, hundreds of which were circulated among the populace during his trial. Mrs. Arnold, in her note, says, 'the General is himself to-day;' meaning that he bore the insults with his usual firmness; but she was alarmed herself, and wished for the presence of some female friend during the painful scene that followed.

"The proof was not enough to condemn Arnold, but there was enough detected of foul play to vitiate his policy. From that time the situation of Arnold at St. John's became even more uncomfortable, and that of his family more distressing. Mrs. Arnold was treated with great kindness, but he was both shunned and despised. She was a lady of great delicacy and refinement, with a mind cultivated with more than ordinary care; and, of course, her sufferings were rendered acute by the imputations against her husband's integrity, aside from his treason. They shortly left St. John's and went to England, where Arnold became lost to the public eye, and died in degradation and obscurity in London, June 14th, 1801, sixty-one years of age."

Connected with the history of this house, we give the following extract from the pen of a writer in the New Jersey *Telegraph*.* Whether "Miss Mary Phillipse" was the first love of Washington, we know

---

* We have recently been informed that it is from the accomplished pen of Gen. George P. Morris.

not; nor have we ever before seen it stated by any writer. The historical events related are true, and the filling up, if from the writer's fancy, is interesting as connected with Revolutionary scenes:

"WASHINGTON IN LOVE.

"In 1756—twenty years before the brilliant era which shines like a rich gem in the pages of the world's history—a gentleman named Beverly Robinson occupied a dwelling (situate in New York), which, at that time, was considered a model of elegance and comfort, although, according to the prevailing taste of the present day, it was nothing of the kind. It was standing, very little altered from its original condition, six years ago, on this side of the Hudson river, within two or three miles of West Point. Mr. Robinson enjoyed all the luxuries known to the colony, and some beside, which the other colonists did not know —for instance, a rich and massive silver tea urn, said, by the gentleman's descendants, to be the first article of this kind, and for a long time the only one, used in this country. In this dwelling, so much admired, the space between the floors and ceiling was exceedingly low, and in many of the rooms (set off, about the fire-places, by polished tiles) the rafters were massive and uncovered; and all things else in the structure were exceedingly primitive. In this house were born or reared a brood of the most prominent and inveterate foes to the patriots of the American Revolution, and the object of that struggle, that history mentions. Two generations of the Robinson family bore arms and held offices in the armies of the English King, and fought determinedly against our sires and grandsires.

"Well, in this house—which will have already attached itself to the interest of the reader—the only victory that was ever gained over George Washington took place.

"In 1756, Colonel George Washington of Virginia, a large stalwart, well-proportioned gentleman, of the most finished deportment and careful exterior; a handsome, imposing, ceremonious, and grave personage—visited his firm and much-esteemed friend, Beverly Robinson, and announced his intention of remaining his guest for many weeks. A grinning negro attendant, called Zeph, was ordered to bring in his master's portmanteau, additional fuel was cast into the broad and cheerful fireplace, an extra bottle of prime old Madeira was placed upon the table, whose griffin feet seemed almost twice the original size at the prospect of an increase of social hilarity, and Colonel Washington was duly installed as a choice claimant of old-fashioned and unrestrained hospitality. Seated with Mr. and Mrs. Robinson, overwhelmed with attention, and in possession of every comfort, the visitor evinced unquiet and dissatisfaction. Every sound of an opening or closing door aroused him from apathy, into which he relapsed when it was ascertained that no one was about to enter the apartment. His uneasiness was so apparent that his host at last endeavored to rally him, but without effect. Mrs. Robinson finally came to the rescue, and addressed the Colonel in direct terms.

"'Pray, friend Washington, may we be made acquainted with the cause of your dullness?—There is some reason for it, and that reason lies with us. Tell it.'

"In vain the Colonel argued that nothing had occurred to vex him—that he was not in want of any farther inducement to present or future happiness. His entertainers would not regard his words, but continued their pertinacious endeavors to solve his mystery. At length, wearied by importunity, Washington—then twenty years before his greatness—leaned over the table, played with his glass, attempted to look unconcerned, and whispered to Mr. Robinson the single word, 'Mary.'

"'Yes?' responded Mr. R. interrogatingly, as if unable to comprehend Washington's meaning.

"'Is she well? Does she still abide with you?'

"'She does,' replied the lady of the mansion.

"Washington again became apathetic and contemplative, while several significant glances passed between the gentleman and his wife. Some five minutes were spent in perfect silence, which was only interrupted by the exit of Mrs. R. from the apartment; she speedily returned, accompanied by a beautiful young lady, whom Washington, with a countenance beaming joyfully, arose to greet with becoming respect.

"The young lady was Mary Phillipse, sister of Mrs. Robinson, and daughter of the owner of the Phillipse' estate.

"It was, perhaps, singular; but the time of her appearance and the period of the return of Washington's cordiality, was incidental. Strange as it was, too, midnight found this young lady and Virginian Colonel alone, and in deep conversation. The conjugal twain who had kept them company in the early part of the evening, had retired to their bed-chamber. More remarkable than all, day-light found this company still together. The candles were burned down to the sockets of the sticks, and the fire-place, instead of exhibiting a cheerful blaze, harbored only a gigantic heap of ashes and a few embers. What could have prolonged that interview! Not mutual love; for the parties preserved a ceremonious distance, and the young lady evinced a *hauteur* that could be matched only by her companion in after years. And yet, the truth must be told. There was love on one side; the Colonel, smitten by the graces and rare accomplishments of a lady as beautiful as nature's rarest works, was endeavoring to win her heart, in exchange for his own. He made his confession just as the cold grey of the dawn of the morning broke up the dark clouds in the east. He confessed, in cautious and measured terms, it is true, the extent of his passion, and avowed what it was his earnest hope would be the result; that it was the gain of her heart. The lady hesitated. Was it the modesty of the maiden who dares not trust her lips with the confession of affection it is her heart's desire to make? She respected, although she did not love her interlocutor, and she felt diffident in making known to him the true state of her feelings. At last, candor triumphed over delicacy; she informed Washington, in set terms, that she loved another! She refused him! The greatest of modern men was vanquished, and by a woman! He was speechless and powerless.

"Trembling with compressed lips, and a countenance ashy pale, he crept from the place just as the old negress of the household entered to make preparation for the breakfast. He sought his room, threw himself upon his couch, dressed as he was, and lapsed into a troubled sleep. The only *victory* ever won at his expense penetrated him to the soul. He was unhappy—supremely wretched! The future conqueror of thousands of brave men suffered because he had been rejected by a female. This was his first but not his last wooing.

"Years rolled on upon the mighty tide of time. George Washington was commander-in-chief of the American forces opposed to the royal government. The friend of his early manhood, Beverly Robinson, was the Colonel of the Loyal American regiment raised in this State, and his son was the Lieutenant Colonel. The house we have spoken of was in possession of the 'rebels,' and was occupied by Arnold, the traitor. It was afterwards the temporary residence of Washington.* At the same time the husband of Miss Mary Phillipse, Roger Morris, was a prominent tory, and a member of the council of the colony. Few of the parties were occupied by any reflections of an amorous nature. Time, in its progress, had worked mutations which had severed the closest ties, both of friendship and consanguinity. Those who were most intimate previous to the commencement of the war, were studied strangers, with drawn swords at each others' breasts. Even sons and fathers were estranged and arrayed in opposite ranks—even the child of that illustrious statesman, Dr. Franklin, was a bitter and uncompromising tory. It must not be supposed that the loyalist friends of Colonel George Washington shared any better fate, so far as the acquaintanceship of the Father of his Country was concerned, than others. His old Hudson River friends he had not seen for years. The husband of Mary Phillipse was personally unknown to him. Beverly Robinson, grown grey and care-worn, would scarcely have been recognized.

"Andre was taken and condemned to death, and while under General Woodhull's charge, was visited by Mr. Robinson in the capacity of a species of a commissioner, which protected his person. What was the surprise of Washington, a few days before the time of the execution, to receive a letter from his old friend and entertainer, referring to past events, and claiming on a score of reminiscence, a secret or private interview. The claim was acknowledged, and late at night Mr. Robinson, accompanied by a figure closely muffled in a cloak, was admitted to the General's

* Before Sir Henry Clinton, or any other person knew of Arnold's defection and Andre's projects, Beverly Robinson was in possession of all the facts. A great grandson of his now practices law, or did, not long ago, in this city.

apartment. For a moment these two men, their position so widely different, gazed at each other in silence.

"Recollections of days gone by, of happy days uncorroded by cankering care, prevailed, and they abruptly embraced. Washington was the first to recover his self-possession. Suddenly disengaging himself, he stood erect and clothed in that unequalled dignity which was his attribute, and said:

"'Now, sir, your business.'

"'Is,' replied Robinson, in a choking voice, 'to plead for Andre.'

"'You have already been advised of my final determination,' replied Washington, sternly.

"'Will nothing avail?' asked Robinson, in smothered accents.

"'Nothing! Were he my own son he should pay the penalty due his offence. I know all you will say: you will speak of his virtues; his sisters; his rank, and of extenuating circumstances; perhaps endeavor to convince me of his innocence.'

"Robinson struggled with his emotions a few seconds, but unable to repress his feelings, he spoke but a single word, with such a thrilling accent that he started at the sound of his own voice. That word was George!

"'General Washington, Colonel Robinson,' responded the great patriot, laying great stress on each military title.

"'Enough,' said the other. 'I have one more argument; if that fails me I have done. Behold my friend!'

"'Your friend! Who is he? What is his name?'

"One other single word was spoken as the heavy cloak, in which the mysterious friend was clothed, fell to the floor and exposed the mature figure of Mrs. Morris, and that word, uttered with a start by Washington, was *Mary!* The suspense was painful but brief.

"'Sir,' said Washington, instantly recovering, 'this trifling is beneath your station and my dignity. I regret that you must go back to Sir Henry Clinton with the intelligence that your best intercession has failed. See that these persons are conducted beyond the lines in safety,' continued he, throwing open the door of the apartment, and addressing one of his aids.

"Abashed and mortified, Mr. Robinson and his sister-in-law* took their leave. The woman had gained a conquest once, but her second assault was aimed at a breast invulnerable."

The following is an extract of a letter from a gentleman, dated Tappan, October 2, 1780, and published in the Boston *Gazette* under date of October 16, of that year, "detailing the villany of Arnold and the capture of the unfortunate Andre. It furnishes a good and interesting account of that remarkable and critical incident in the scenes of the war of the Revolution."

"You will have heard before you can receive this, of the *infernal villany* of Arnold. It is not possible for human nature to receive a greater quantity of guilt than he possesses: perhaps there is not a single obligation, moral or divine, but that he has broken through. It is discovered now, that in his most early infancy, hell marked him for her own, and infused into him a full proportion of her diabolical malice.

"His late apostacy is the summit of his character. He began his negotiations with the enemy, to deliver up West Point to them, long before he was invested with the command of it, and whilst he was still in Philadelphia; after which he solicited the command of that post, for the ostensible reason, that the wound in his leg incapacitated him from an active command in the field. It was granted to him on the 6th of August last.

"Since which he has been as assiduous as possible in ripening his plans, but the various positions the army assumed, prevented their being put into execution.

"On the night of the 21st ultimo, he had an interview with Major Andre, the Adjutant-General of the British Army. This gentleman came on shore from the Vulture man-of-war, which lay not far from Teller's Point, to a place on the banks of the river, near to the Haverstraw mountain, where he met Arnold,

---

* "Her husband had been an aid of Braddock, and had been the companion in arms of General Washington."

who conducted him to the house of Joshua Smith (the white house), within our lines, and only two miles from Stony Point. They arrived in the house just before day, and stayed there until the next evening, when Major Andre became extremely solicitous to return by the way he came, but that was impossible, for the two men whom Arnold and Smith had *seduced* to bring on shore, refused to bring him back. It then was absolutely necessary he should return to New York by land. He changed his dress and name, and thus disguised, passed our post of Stony and Verplank's Points, on the evening of the 22nd ult., in company with the said Joshua Smith, brother to William Smith, Esq., Chief Justice within the British lines; he lodged that night at Crom Pond, with Smith, and in the morning left Smith, and took the road to Tarry Town, where he was taken by some militia lads about 15 miles from King's bridge. He offered them any sum of money, and goods, if they would permit him to escape, but they readily declared and inflexibly adhered to it, that 10,000 guineas, or any other sum, would be no temptation to them. It was by this virtue, as glorious to America as Arnold's apostacy is disgraceful, that his abominable crimes were discovered.

"The lads in searching him, found concealed under his stockings, in his boots, papers of the highest importance, viz.:—

"1. Returns of the ordnance and its distributions at West Point and its dependencies.

"2. Artillery orders, in case of an alarm.

"3. Returns of the number of men necessary to man the works at West Point, and its dependencies.

"4. Remarks on the works at West Point, with the strength and working of each.

"5. Returns on the troops at West Point, and their distributions.

"6. State of our army, &c., transmitted by General Washington to Arnold, for his opinion, which state had been submitted to all the general officers in the camp, for their opinions.

"Besides which it appears, that Arnold had carried with him to the interview, a general plan of West Point and its vicinity, and all the works, and also particular plans of each work on a large scale, all elegantly drawn by the engineer at that post.

But these were not delivered to Major Andre, and from their requiring much time to copy, it is supposed they were not to be delivered until some future period.

"From some circumstances, it appears that it was not Arnold's intention to have deserted, but that he meant to be taken at his post, which, from his distribution of the troops, it was very easy to have seized.

"His Excellency, the General, on his return to camp, determined to visit West Point, and in pursuance of that plan, was viewing some redoubts which lay in his way to Arnold's quarters. He had sent out servants there, and Major Shaw and Dr. McHenry had arrived, and were at breakfast with the traitor when he received intelligence by letter of Andre's being taken. His confusion was visible, but no person could divine the cause. He hurried to his barge with the utmost precipitation, after having left word that 'he was going over to West Point and should be back immediately.' This was about ten in the morning of the 25th ultimo.

"The General proceeded to view the works, wondering where Arnold should be; but about 4 o'clock in the afternoon he was undeceived, by an express with the papers taken on Andre. The apostate at this time was on board the Vulture, which lay about five or six miles below Stony and Verplank's Points.

"Major Andre was brought to the General at West Point, and from thence he was brought to this camp. A board of general officers have examined into his case, and upon his own most candid confession, were of opinion that he was a spy, and according to the custom and usages of nations, he ought to suffer death; and about two hours ago he was executed.

"This gentleman was in the highest degree of reputation in the British army, of the most polite and accomplished manners, extremely beloved by Sir Henry Clinton. His deportment while a prisoner was candid and dignified. He requested no favor, but to die the death of a soldier, and not on a gibbet. Rigorous policy forbade granting a favor which at first flash seems immaterial. Our army sympathizes in the misfortunes of the Chesterfield of the day. But if he possessed a portion of the blood of all the Kings on earth, justice and policy would have dictated his death.

"The enemy, from hints that some of the officers dropped, appeared to be inclined to deliver Arnold into our hands for Major Andre. But they since declared it was impossible. If it could have been effected, our desire to get Arnold would have rendered the exchange easy on our part.

"The British army are in the utmost affliction on the account of Major Andre, and have sent repeated flags on the subject. Yesterday they sent General Robertson, Andrew Elliot, and William Smith, Esqrs.—the two latter were not permitted to land. General Green met General Robertson; he had nothing material to urge—' but that Andre had come on shore under the sanction of a flag, and therefore could not be considered as a spy;' but this is not true, for he came at night, had no flag, and on business totally incompatible with the nature of a flag. He also said they should retaliate on some people at New York and Charlestown; but he was told that such conversation could neither be heard nor understood. After which, he urged the release of Andre on motives of humanity, and because Sir Henry Clinton was much attached to him; and reasons equally absurd.

"I have been particular in this narration, well knowing what strange stories you will have on the subject."

The following is a copy of a letter from Major Andre to His Excellency Gen. Washington:

"Sir,—Buoyed above the fear of death, by the consciousness of a life spent in the pursuit of honor, and fully sensible that it has at no time been stained by any action, which at this serious moment could give me remorse—I have to solicit your Excellency, if there is anything in my character which excites your esteem; if aught in my circumstances can impress you with compassion; that I may be permitted to die the death of a soldier.—It is my last request, and I hope it will be granted.

"I have the honor to be," &c.

"MAJOR ANDRE'S DEFENCE.

"A correspondent of the New York *Daily Advertiser*, who seems to be fortunately in the possession of sundry curious old papers and other memorials of the past, as well as of corresponding knowledge and memory, has furnished for the columns of that paper a document which we do not remember to have ever seen before—the defence read by Major Andre before the Court which condemned him to death as a spy. We have no doubt that it will be read with lively interest by many:

"'I came,' he said, 'to hold a communication with a general officer of the American army, by the order of my own commander. I entered the American lines by an unquestionable authority—when I passed from them it was by the same authority. I used no deception. I had heard that a provincial officer had repented of the course he had taken, and that he avowed that he never meant to go as far as he had gone, in resisting the authority of his king.

"'The British commander was willing to extend to him the King's clemency—yea, his bounty—in hopes to allure others to do the same. I made no plans. I examined no works. I only received his communications, and was on my way to return to the army, and to make known all that I had learned from a general officer in your camp. Is this the office of a spy? I never should have acted in that light, and what I have done is not in the nature of a spy. I have noted neither your strength nor weakness. If there be wrong in the transaction, it is mine? The office of a spy a soldier has the right to refuse; but, to carry and fetch communications with another army, I never heard was criminal. The circumstances which followed after my interview with General Arnold, were not in my power to control. He alone had the management of them.

"'It is said that I rode in disguise. I rode for security *incog.*, as far as I was able, but other than criminal deeds induced me to do this. I was not bound to wear my uniform any longer than it was expedient or politic. I *scorn* the name of a spy; brand my offence with some other title, if it change not my punishment, I beseech you. It is not death I fear. I am buoyed above it by a consciousness of having intended to discharge my duty in an honorable manner.

"'Plans, it is said, were found with me. This is true; but they were not mine. Yet I must tell you honestly that they would have been communicated if I had not been taken. They were sent by General Arnold to the British Commander, and I should have delivered them. From the bottom of my heart I spurn the thought of attempting to screen myself by criminating another; but so far as I am concerned, the truth shall be told, whoever suffers. It was the allegiance of General Arnold I came out to secure. It was fair to presume that many a brave officer would be glad at this time to be able to retrace his steps; at least, we have been so informed. Shall I, who came out to negotiate this allegiance only, be treated as one who came to spy out the weakness of a camp? If these actions are alike, I have to learn my moral code anew.

"'Gentlemen, officers, be it understood that I am no suppliant for mercy; *that* I ask only from Omnipotence—not from human beings. Justice is all I claim—that justice which is neither swayed by prejudice, nor distorted by passion, but that which flows from honorable minds, directed by virtuous determinations. I hear, gentlemen, that my case is likened to that of Capt. Hale, 1775. I have heard of him and his misfortunes. I wish that in all that dignifies men, that adorns and elevates human nature, I could be named with that accomplished but unfortunate officer. His fate was wayward, and untimely was he cut off, yet younger than I now am. He went out, knowing that he was assuming the character of a spy. He took all its liabilities into his hand, at the request of his great commander. He was ready to meet what he assumed, and all its consequences. His death the law of nations sanctioned. It may be complimentary to compare me with him, still it would be unjust. He took his life in his hand when he assumed the character and the disguise. I assumed no disguise, nor took upon myself any other character than that of a British officer who had business to transact with an American officer.

"'In fine, I ask not even for justice; if you want a victim to the manes of those fallen untimely, I may as well be that victim as another. I have, in the most undisguised manner, given you every fact in the case. I only rely on the proper construction of those facts. Let me be called anything but a spy. I am not a spy. I have examined nothing, learned nothing, communicated nothing, but my detention, to Arnold, that he might escape if he thought proper so to do. This was, as I conceived, my duty. I hope the gallant officer, who was then unsuspicious of his general, will not be condemned for the military error he committed.

"'I farther state that Smith, who was the medium of communication, did not know any part of our conference, except that there was necessity for secrecy. He was counsel in various matters for General Arnold, and *from* all the interviews I had with him; and it was Smith who lent me this dress-coat of crimson, on being told that I did not wish to be known by English or Americans. I do not believe that he had even a suspicion of my errand. On me your wrath should fall, if on any one. I know your affairs look gloomy; but that is no reason why I should be sacrificed. My death can do your cause no good. Millions of friends to your struggle in England, you will lose, if you condemn me. I say not this by way of threat: for I know brave men are not awed by them—nor will brave men be vindictive because they are desponding. I should not have said a word had it not been for the opinion of others, which I am bound to respect.

"'The sentence you this day pronounce will go down to posterity with exceeding great distinctness on the page of history; and if humanity and honor mark this day's decision, your names, each and all of you, will be remembered by both nations when they have grown greater and more powerful than they now are. But, if misfortune befals me, I shall in time have all due honors paid to my memory. The martyr is kept in remembrance when the tribunal that condemned him is forgotten. I trust this honorable Court believes me, when I say that what I have spoken was from no idle fears of a coward. I have done.'"

The following recapitulation of the judgment of the Court Martial before whom Major Andre was tried, the order by Washington approving the same, and directing its execution, is taken from the "Revolutionary Orders" of the Commander-in-Chief, edited by Henry Whiting, Lieut. Col. U. S. Army, from the manuscripts of his father, John Whiting, Lieut. and Adjutant of the 2d Regt. Mass. Line:

"No 80. Head Quarters, Orange Town, October 1st, 1780.

"The Board of General Officers* appointed to examine into the case of Major Andre, have reported—1st, That he came on shore from the Vulture Sloop of War in the night of the 21st of September last, on an interview with General Arnold, in a private and secret manner; 2ndly, That he changed his dress within our lines, and under a feigned name and disguis'd habit, pass'd our works at Stoney and Verplank's Points the evening of the 22nd of September last, and was taken the 23rd of September last, at Tarrytown, in a disguis'd habit, being then on his way to N. York, and when taken, he had in his possession several papers which contained intelligence for the Enemy.

"The Board having maturely consider'd those facts, do also report to his Excellency, General Washington, that Major Andre, Adjutant-General of the British Army, ought to be considered as a Spy from the Enemy, and that, agreeably to the law and usage of Nations, it is their opinion that he ought to suffer death. The Commander-in-Chief directs the execution of the above sentence, in the usual way, this afternoon, at 5 o'clock."†

Beverly Dock.—This dock was built by Beverly Robinson, whose christian name it bears, soon after erecting his mansion. Originally it was about twenty feet long from east to west, and ten feet wide from north to south. It is built against the base of a small rocky promontory projecting in a southerly direction into the Hudson, between which and the high-grounds on the east, is a small, narrow swale, covered with long, rank water-grass. The mountain-road leading

* "The Board referred to consisted of Major-Gen. Greene, as President, and Major-Generals Marquis La Fayette and Baron Steuben."

† "In the 'After General Orders,' it was announced that 'the execution of Major Andre is postponed till to-morrow.' In the 'Evening Orders,' of the same date, it was announced, 'Major Andre is to be executed to-morrow at 12 o'clock precisely. A battalion of eighty files from each wing to attend the execution.'"

to this dock, alluded to by different writers, commences about fifty rods north of the "Robinson House," on the west side of the road, and runs in a south-westerly direction, crossing the high and rocky grounds near the river, and descending into the north end of the swale, winds along the eastern base of the rocky barrier, separating it from the river, to the "Dock." A section of the Hudson River Railroad is now being cut through this rocky barrier, nearly grazing the eastern side of the "Dock." This is "the mountain road Gen. Washington and his suite were walking up from Beverly Dock," when General Hamilton met him, and taking him to one side, briefly informed him of Arnold's treason, the undoubted evidence of which, in his own handwriting, had reached "Beverly," that morning, after Washington's departure to West Point to see the *Traitor*.

We have followed the track he is said to have taken on his departure for the "Dock." About two rods south of the new corn-house built by Lieut. Arden, the accomplished and gentlemanly occupant of the premises, "whence no visiter departs, who can ever forget the generous Highland welcome," was a gate leading into the cleared field; through this Arnold dashed, and crossing the field in the direction of the river, he passed through a second gate on its western side, entering the woods on the brow of a very steep and abrupt descent, and plunging down it on a gallop, he came into the mountain-road, a few rods north of the "Dock." His horse must have been as sure-footed as that of Putnam's, when he descended the steep hill at Horse Neck, to have carried him safely to the bottom.

From this "Dock" Arnold entered his barge and departed from the Highlands never more to return. Standing upon the deck of some of the Hudson River steamers, we have often passed this little spot of revolutionary ground, and witnessed groups of travellers surveying the eastern shore of the Hudson with the *Traveller's Guide Book* in hand, eagerly inquiring, "Which is the 'Beverly Dock?'" "*Where is the spot* where the *Traitor* took boat for the *Vulture?*"

In a Plymouth paper, in July, 1825, appeared the following notice of an application for a pension by one of Arnold's bargemen, detailing the manner of his departure from the "Beverly Dock," and copied by the Hon. S. W. Eager, in his History of Orange County, from which we extract it:

"Application was made this week in this town for assistance in making out the necessary documents for a pension by one of the bargemen in the barge that conveyed Gen. Arnold to the sloop of war Vulture. He was bow-oarsman in the boat, next in rank to the coxswain, whose name was James Larvey. His memory is remarkably accurate, and his veracity is unquestionable. He is a brother to Mr. James Collins of this town. The day before the flight of Arnold, he brought him with Major Andre, from Lawyer Smith's, below Stony Point, to the General's head qnarters. They conversed very little during the passage. The General told his aid, who was at the landing when they arrived, that he had brought up a relation of his wife. Arnold kept one of his horses constantly caparisoned at the door of his quarters, and the next morning soon after breakfast he rode down in great haste with the coxswain just behind him on foot. The coxswain cried out to the bargemen to come out from their quarters, which were hard-by, and the General dashed down the footpath, instead of taking a circuit, the usual one for those who were mounted. The barge was soon made ready, though the General, in his impatience, repeatedly ordered the bow-man to push off, before all the men had mustered. The saddle and holsters were taken on board the barge, and Arnold, immediately after they had pushed off, wiped the priming from the pistols, and primed anew, cocked and half-cocked them repeatedly. He inquired of Collins if the men had their arms, and was told that the men came in such haste, that there were but two swords belonging to himself and the coxswain. They ought to have brought their arms, he said. He tied a white handkerchief to the end of his cane for a flag in passing the forts. On arriving alongside of the Vulture he took it off and wiped his face. The General had been down in the cabin about an hour when the coxswain was sent for, and by the significant look and laughing of the officers, the men in the barge began to be very apprehensive that all was not right. He very soon returned, and told them that they were all prisoners of war. The bargemen were unmoved and submitted, as to the fortune of war, except two Englishmen, who had deserted, and who were much terrified, and wept.

"The bargemen were promised good fare if they would enter on board the Vulture, but they declined and were handcuffed, and so remained for four days. Gen. Arnold then sent for them at New York. In passing from the wharf to his headquarters, the two Englishmen slipped aboard a letter of marque, then nearly ready to sail. The others, five in number, waited on Arnold, who told them they had always been attentive and faithful, and he expected they would stay with him. He had, he said, command of a regiment of horse, and Larvey, you and Collins may have commissions, and the rest shall be non-commissioned officers. Larvey answered that he could not be contented—he would rather be a soldier where he was contented, than an officer where he was not. The others expressed or manifested their concurrence in Larvey's opinion. He then gave the coxswain a guinea, and told them they should be sent back. At midnight they were conveyed to the Vulture, and next day sent on shore. This worthy and intelligent applicant perfectly remembers Major Andre's dress, when they took him up in the barge, from Smith's house to Arnold's quarters—blue homespun stockings—a pair of wrinkled boots, not lately brushed—blue cloth breeches, tied at the knee with strings—waistcoat of the same—blue surtout, buttoned by a single button—black silk handkerchief once round the neck and tied in front, with the ends under the waistcoat, and a flapped hat."

It was in the beginning of August, 1780, that Arnold arrived at West Point, and established his head-quarters at the "Robinson House." Washington arrived there about the middle of September, and tarried a few days inspecting the posts, while the Vulture lay at anchor in the river below. In crossing one of the ferries with Washington and his Staff, the vessel was seen at a distance, having on board, as Arnold well knew, Colonel Robinson, sent by Sir Henry Clinton to meet him. Washington watched the vessel with his glass, whilst Lafayette jocularly remarked, that Arnold ought to find out what had become of the expected naval reinforcements from France, as he had convenient modes of intercourse with the enemy. For a moment Arnold lost his presence of mind, and made a reply, the intemperance of which might have roused suspicions of any other man. But Washington entertained none, and the matter dropped. The next day (19th September) Washington continued his journey to Hartford, and Arnold was left to his unimpeded work of villany. His first step was to advise Sir Henry Clinton that he would be in attendance under due precautions, the next day, near Dobb's Ferry, ready to meet his messenger. The following hurried letter to a forage agent in the neighborhood, has never before been published, and bears date the day that Washington and Arnold parted. The autograph indicates hurry and agitation :

"To Mr. Jefferson, Fredericksburg, N. Y.

"Headquarters, Rob. House, September 19th, 1780.

"Sir—You will please to pick out of the horses you have now in your custody, or which you may hereafter receive, a pair of the best wagon horses, as also two of the *very best saddle horses* you can find for my use. You'll send them to me as soon as possible.

"I am, Sir, your most obedient servant,
"B. Arnold, M. General."*

REVOLUTIONARY HOUSES STILL STANDING.

1. The "Robinson House" built by Col. Beverly Robinson, now owned by Richard D. Arden, Esq. In the Revolution, it was the *Head Quarters* of Arnold, Gen. Heath, and others who had charge of West Point and its out-posts.

2. The old house now occupied by George Haight, about one mile south-east of Philips' paper-mill, on the road leading from John Garrison's to the Peekskill turnpike, near Henry Croft's. It was built by Daniel Haight, father of the present occupant.

3. The old house where Cornelius Haight now lives, on the old post-road, about a mile south of Nelson's mill. It was built by John Rogers, and occupied by him during the French and Revolutionary war.

4. The house where James Croft now lives. It was occupied in the Revolution by James Croft, father of Henry Croft, Esq.

5. The old house now occupied by Edward Hopper. It was built in 1772, by Richard, father to the present Edward Hopper, and occupied by him during the Revolution.

---

* The original of this letter is in possession of Edward D. Ingraham, Esq., of Philadelphia.

6. The old house now occupied by John Mills Brown, on the old road leading from Continental Village to Bross's Landing, about one fourth of a mile below the mansion house of the Hon. John Garrison.

7. The old house now occupied by John Hopper, on the old post-road, on the hill south of Nelson's mill. It was occupied during the Revolution, by Samuel Warren, grandfather of the Hon. Cornelius and Sylvanus Warren.

8. The old house now occupied and owned by John Griffin, Esq., about three miles from Cold Spring, on the turnpike leading to Mill Town. It was built by Gilbert Budd, who lived there in 1750, and during the Revolution.

9. The old house where Thomas Jaycox recently lived, on the turnpike about two miles east of John Griffin's tavern. It was owned by James Jaycox.

10. The house now occupied by the widow Miller. In the Revolution, it was occupied by Isaac Garrison.

11. The old house where Jacob Denike lives. It was then occupied by Jacob Denike, deceased.

SILVER AND LEAD MINES.

By the mouth of tradition, it is asserted that a silver mine was discovered in this town, as far back as 1763. A man by the name of Jubar, coined money; and it began to be rumored, that he obtained the ore in this town. An examination was set on foot by the King's government and the fact established, that the money coined by Jubar contained silver, mixed with other metals, but not in sufficient quantity to deceive a practised eye. If he had discovered a mine in this town, in which silver was mixed with other metals, its location was never known, but tradition places it in the neighborhood of the Sunk Lot. He was arrested by order of the Colonial Government, tried at Poughkeepsie, found guilty, and hung about the year 1765. A man by the name of Samuel Taylor, was associated with him; and on the arrest of Jubar, left this part of the country. He returned in the Revolution, or shortly after its close, became poor, and died "on the town." He always said that Jubar melted an ore, from which he extracted silver.

According to tradition, the next discovery of a silver mine in this town, was made three or four years before the Revolution, by one Eleazer Gray, a silversmith by trade, who lived in the middle of the Sunk Lot. His father, John Gray, had a gristmill in the Revolution, a short distance above Bunell's forge on the Sunk Lot. Eleazer put up a log shop to work the ore in. Squire Peterson and Beverly Robinson, hearing that the younger Gray had discovered a silver mine and was melting ore, went to his house for the purpose of ascertaining the truth of the reports. Robinson, who was somewhat inclined to believe that he had discovered a mine, said to him, "Gray put up a shop by your house, and not work underhanded in the swamps, and you may have all the profits of it." But no bribe could tempt him to reveal what he professed to know respecting the mine.

In 1780, Edward Hopper, now living, went to John Gray's mill to get some grain ground. Gray told him that the water was low, and if he could wait till near night, he would then be able to grind his grain. Gray had a younger son than Eleazer, by the name of John, who proposed to Hopper to go, in the meantime, to the brook above and catch a mess of trout. The water being low, they could catch them with their hands. They left the mill and went a mile, and when near a swamp, young Gray said to Hopper, "*Did you ever hear* anything of Eleazer's mine?" "Yes," replied Hopper. "We are close to it now," continued Gray. They came to a spot where a considerable quantity of rocks had been blown out. Mr. Hopper, now a very old man, says that there was a ledge of rocks a little west of the place, and a spring of water a few rods south of it. While loitering there, young Gray said to Hopper, "This is the best ore Eleazer ever found." Eight or nine years afterwards Mr. Hopper visited the same spot, and found that the blasted rocks had been removed, and the evidence thereof carefully covered up.

Towards the latter part of the Revolution the Grays, in consequence of a parcel of horse-thieves having been seen at their residence, became suspected, and it was thought they were leagued with those midnight desperadoes. Their counterfeiting operations had leaked out and gave them an unenviable notoriety.

Their neighbors burnt their house, shop, and barn down with the view of inducing them to quit that part of the country.

They then moved down to where James Croft now lives. Shortly afterwards the whole family moved down to Sugar Loaf, in Orange county, where the elder Gray, and his son Eleazer, died.

About the year 1800, a man by the name of Henry Holmes was arrested for counterfeiting metal money in this town. It is said that he carried on his operations in a cave or hole in the rocks near, or in the vicinity of the residence of Richard Denny. Holmes was from Westchester county, and is supposed to have had an accomplice who assisted him. His money contained so little of the precious metals, that the counterfeit was apparent to the most ordinary business man; and when it was thrown against a hard body would break with the facility of a pipe-stem.

Holmes was tried, found guilty, and sentenced to the State Prison for seven years. His supposed accomplice, who made his moulds, was acquitted in consequence of an informality in the indictment.

In 1812, Joseph Parks, Pelick Wixon, and Nathaniel Tompkins, it is said, while engaged in making the Cold Spring turnpike discovered a lead mine, along the route of the road. They all went before Doctor Baily, a magistrate, by previous agreement, and made oath that neither of them would reveal the location of the mine while two of the three should be living. Parks, at that time, was overseer of the road. After the road was built he went to New Orleans, built a turnpike there, and returned to the city of New York, where he died. Tompkins and Parks then sent John Baily, a son of the Doctor, and the Magistrate before whom they mutually made oath, to buy the land. The price was agreed upon between the owner and young Baily; but when the deed was made out, it contained a reservation of all mines and minerals on the land to be conveyed, and the deed was refused.

THE TRIAL OF GEORGE DENNY,

Convicted of the murder of Abraham Wanzer, at the May term of the Putnam Oyer and Terminer in 1844; and his confession made to Benjamin Baily, Esq., his counsel.

"At the October Term of 1843, George Denny, whom it was said was nearly eighteen years of age, but whose appearance denoted him to be much younger, was placed on his trial, charged with the murder of Abraham Wanzer, an old man, nearly eighty years of age, to which he pleaded *not guilty*. The evidence, which was similar to that given on the recent trial, it appears was not satisfactory to the jury, who were discharged, being unable to agree. He was remanded to prison, and at the May Term of said court in 1844, was again put on his trial for the same charge—the Hon. Amassa J. Parker, Judge of the Third Circuit, presiding.

"It was shown on the part of the people by the widow of Mr. Wanzer, that the deceased lived in a retired part of Philipstown, in said county, in a small log-house, on a little spot of ground, which he cultivated—that about a year before the death of deceased, the prisoner returned a key which he acknowledged he had stolen from him—that deceased advised him to be a good boy, and never again be guilty of such a foolish transaction—that on the evening of the ninth of October, 1843, about eleven o'clock, some one raised the latch of their door; she enquired who was there? and was answered a friend; she wished to know what a friend wanted at that time of night, and was replied to that he wished to stay all night; that from the sound of his voice he appeared to be receding from their door during the conversation —that on the next morning, the 10th, she discovered that some one had laid in their barn the night previous—that she thought at the time, and also remarked to her husband, it was the prisoner—that at intervals during the day, she heard the report of guns in the neighborhood of their house—that about seven o'clock the same evening, something like the gnawing of a dog, was heard at their door—the deceased opened the door, and called his little grand-daughter to know if it was not a neighbor's dog—she replied no, it was a black-back dog with short legs—that some one whistled, the dog left, and the deceased resumed his seat—that shortly after, something struck the end of their house, as if a stone had been thrown—that her husband went to ascertain the cause. In a few moments she heard the report of a gun; and on going into their yard found deceased lying on his back dead, and that immediately, with her two little granddaughters, she proceeded to the nearest neighbor, and gave the alarm.

"The eldest of the two grand-daughters corroborated the testimony of Mrs. Wanzer in the main; and on producing the dog of the prisoner in court, she said that it was the same she saw in front of their door on the evening of the death of her grandfather.

"Wm. W. Johnson testified that he was a surgeon, that he examined the body of the deceased—that he found a wound in the left side, near the heart—that there were twenty-seven shot holes, and one bullet hole within a space of two inches diameter—that he extracted from the body two sizes of shot and one ball, corresponding in size and weight with balls handed to him by Francis Booth, which witness produced in court.

"Abraham Knapp testified that he was a brother-in-law of the prisoner; that on the morning of the ninth of October, 1843, the prisoner loaned his two-barrel gun, the dog produced in court, together with a shot bag and powder-horn, for the purpose, as prisoner said, of hunting—that he, witness, cast some balls the spring previous, in a mould which witness got of his father—that some of the balls were in his house a short time before the murder, that prisoner was in the habit of staying at witness's house, and had access to the balls—and that he lived about nine miles from the residence of deceased.

"Lemuel Wixon testified that he, in company with Francis Booth, on the 11th of October, 1843, arrested the prisoner near Abraham Knapp's house, that they found with him the dog in court, a two-barrel gun, and a shot bag, but no powder-horn or balls—that there were three kinds of shot in his possession, and that the prisoner did not manifest any fears, and made no resistance.

"Francis Booth corroborated Wixon, and further testified that he loaned the bullet moulds produced in court, of Abraham Knapp's father, and that he cast the balls he gave to Dr. Johnson in said moulds.

"Thomas Davenport testified that he accompanied the prisoner after his examination to the County jail, that on their way prisoner pointed out a stump about three miles from deceased's, and said that he shot at it on the day that Wanzer was killed in the evening—that he, witness, afterwards found three kinds of shot in said stump, but no ball.

"Marvin Wilson testified that in searching in an oat field on the east and adjoining the premises of deceased, about a week after his death, he discovered tracks made as if by some person in the act of running from deceased's house, that he took measures to preserve them—that within five or six days he visited the jail and found the prisoner's boots and compared them with the tracks—that the boots were rights and lefts—that the heel of the left one was worn off on the inside—that he compared them with the tracks and they precisely fitted, and that the tracks led in the direction of a crossing place over the creek in a path leading from the deceased's to the Turnpike.

"John Garrison testified that on the day of deceased's death, he saw a person about three miles east of his house, near the Turnpike, going into the woods, dressed as the witnesses described the prisoner to have been at the time of his arrest, and having with him a double barrel gun and a little dog.

"Mary Denny testified that on the evening of deceased's death she heard the report of a gun; she lived about half a mile east from his residence; that after she had retired to rest, her dog barked, and on looking out of the window, she saw a person passing east on the Turnpike with something like a staff or gun, and as she also thought a dog, and dressed similar to the prisoner at the time of his arrest.

"Paulina Conklin testified that on the day prisoner was arrested she saw him pass her house in the afternoon, going easterly. Witness lived a little over three miles east from the deceased's.

"Joseph Derbyshire testified that three years before the prisoner told him there were three or four he wanted to shoot, and if Mr. Wanzer did not hush up about the key he would be the first. Witness admitted he had had a difference with prisoner.

"Peter Vantassel heard prisoner say he would have a drop of Wanzer's heart's-blood, but could not tell when he heard it.

"John A. Miller heard prisoner say the spring before, when asked if he had returned the key, that he would fix Wanzer yet.

"Richard Laforce testified that about a month before the death of deceased he was confined with prisoner in Poughkeepsie jail—that prisoner said they had put him there for no good—that he was studying deviltry, and that when he got out there were three or four he meant to shoot. Witness claimed that he was confined for stealing honey in connection with prisoner, but that he was innocent.

"Here the testimony on the part of the people closed. For the following remarks as to the defence, the summing up of Counsel, the charge of the Court, and sentence of the prisoner, the writer is indebted to a friend, as he was not in a situation to report proceedings.

"The prisoner proved in his defence that a person having a gun and small black dog, and dressed similar to himself, but being a somewhat larger person, was seen on Tuesday morning 10th October, a little distance east of Cold Spring, going easterly towards Wanzer's.

"He also showed that a neighbor was at Wanzer's house about sunset on that evening with a gun.

"He proved also, that he stated to a person, before he was arrested, he had been lost in the woods two or three days, without anything but chestnuts or grapes to eat, that he was not afraid, and that he was going to take the gun to Abraham Knapp's. He also proved that he was in the habit of going frequently two or three days without eating much of anything. He also showed by the witnesses for the prosecution that he made no attempt to escape or resist when he was arrested; and on this defence rested his case.

"The evidence was conducted on both sides throughout the whole examination with the greatest perseverance and faithfulness.

"The summing up was commenced by Benjamin Baily, counsel for the prisoner, who in a clear and masterly speech of an hour and three-quarters, discharged his duty in behalf of the prisoner, in a most able, affecting, and faithful manner.

"Frederick Stone, District Attorney, spoke on the part of the prosecution about an hour, briefly illustrating and presenting to the Jury the most material points of evidence against the prisoner. He was followed by Thomas R. Lee, counsel for the prisoner, who, in an able and indefatigable speech of three hours' length, reviewed all the evidence and showed with much clearness the great hazard of relying upon circumstantial evidence, which is generally more or less casual and perfidious.

"William Nelson, Counsel on the part of the People, then closed the summing up of the cause, in a forcible and ingenious speech of two hours' and three-quarters length.

"His Honor, Judge Parker charged the jury in a very able and eloquent manner, clearly presenting to their view all the most prominent and material points of the evidence given on the trial, which were calculated, either to criminate or to show the nnocence of the prisoner.

"After receiving their charge, which occupied about an hour, the jury retired to their room, and in about four hours returned with a verdict of guilty.

"On the last day of term the prisoner was brought in Court to hear his sentence. The Court ordered him to stand up, and asked him if he had anything to say why sentence should not be passed upon him.

"He replied that he had nothing to say.

"The Court then said to him, they must proceed to perform the most painful duty required by law—that he had had a fair and impartial trial—that the counsel who were engaged for him, had defended his cause with the greatest care and faithfulness, which was creditable to themselves and honorable to their profession—that there was no doubt of his guilt—that the evidence clearly established it, and that a jury had pronounced him guilty—that Abraham Wanzer, a peaceable citizen, living with his family in a retired spot, without a known enemy, respected by all who knew him, and advanced in old age, was called from his dwelling and shot without the least provocation—that such an act was sufficient to bring down the just indignation of every citizen upon the perpetrator of the deed—that it was their duty to search out the offender and inflict the full penalty of the law upon the wretch who could thus deliberately take the life of an inoffensive fellow-being—that it was extremely painful and grievous to see the great indifference and unconcern which he had manifested throughout the whole course of his imprisonment and trials—that owing to his age and condition the Court would give him the longest time before his execution, that the law would allow in his sentence—that he ought to take the whole time to reflect, and prepare himself for the judgment of another world—that there was an All-Seeing eye who saw him commit the deed—that he must not entertain any hopes of a pardon—that he ought to make a full confession of his guilt, as this was the first step towards repentance—that he ought sincerely to repent and prepare for the awful change that so soon awaited him—that he might yet find pardon before the Judge of a higher world.

"With these remarks, which from recollection we have briefly attempted to sketch, the Court sentenced him to be hanged by the neck on the 26th day of July next, between 5 o'clock A. M. and 7 o'clock P. M., until dead."

"CONFESSION OF GEORGE DENNY.

"To the Hon. AMASA J. PARKER,

"Dear Sir:—A few days after the sentence of the above named most singular and unfortunate individual, I had an interview with him. I found him engaged in reading his Bible. Having said that I was pleased to see him endeavoring to prepare for the awful doom which awaited him, and from which I presumed there was not the most distant hope of escaping, I requested him, as one who had taken a strong interest in his favor up to the time of his conviction, and who desired to lend his best efforts towards ensuring him peace in his last moments, to frankly and fully give me the history of himself, and especially of his participation in the crime of which he stood convicted. As I anticipated, he declined. After conversing with him somewhat fur ther, he requested me to explain the meaning of the 22d verse of the 14th chapter of Romans. I gave it as my opinion that the passage to which he referred me was no obstacle to a confession; and if he desired to place himself in a situation to understand what he read, it would be necessary for him to free his mind of the heavy burden, which I had every reason to believe was resting upon it, by giving a full and perfect statement of the whole transaction. He at length consented, and gave it as follows:

"I am about 18 years of age; I was born in Putnam County. My mother died when I was an infant, as I am informed, insane. My father abandoned his wife and children a short time before my birth, unprovided for and unprotected. When I arrived to the age of eight years he returned and took with him my sister and myself to the State of Michigan, where we remained about one year, during which time he was convicted and sentenced to prison for two years, for robbing a store. He made his escape by digging under the walls, and returned to his suffering children. Within a few days he was retaken and imprisoned. My sister about fourteen, and myself about eight years of age, without friends or necessary means, after many hardships, returned to my grandfather's in this county. That sister, from my infancy up to this moment, has been my warmest, and I can almost say my only friend, she has often given me good advice, and it is my earnest prayer that she may yet be rewarded. Here I would say to parents, and to all who have the charge of children, cultivate in them habits of industry and honesty, as I have every reason to believe, if my mind had been turned into the proper channel in my infancy, I should not be where I am. Two or three years after our return from Michigan, my father visited us and remained about one month. His mind appeared to have undergone a sad change since we had last seen him. He published a pamphlet founded on the book of Revelations, in which he represented himself as Jesus Christ. I recollect of his saying to my grandfather one day, that Buffalo was the promised land—that he should assemble all the people there, and amongst the number the Queen of England. I stepped up and told him he was a damned fool. He became very much enraged and pursued me out of the meadow, but I got out of his reach. The first inquiry he made of me was, 'George, are you old enough to handle the sword?' He left, and we have never heard from him since.

"During the time I lived with my grandfather I had an opportunity to attend school, but having the privilege of doing as I pleased I seldom attended—my attendance at church came under the above rule. My grandmother indulged me in every evil habit, and my education in consequence is very limited. I can make out to read by spelling some of the words, but cannot write. When I pilfered money from my grandfather, which was not unusual, I was sure to find protection by appealing to my grandmother. With the boys of the neighborhood I bore the appellation of 'the cunning little thief,' and many times have I been reproached and called a fool by some of my relations for acknowledging my thefts, which was usually the case if I was accused. With the exception of some trifles, and the money I took from my grandfather and Mr. Wanzer's key, the first I ever stole was $4,75, from Andrew Miller's trunk, which I opened with Mr. Wanzer's key. I went on from one petty theft to another until I was compelled to leave my grandfather's for fear of an arrest, when I found my way to Shanandoah in the town of Fishkill. I remained there the better part of a year, sleeping in the barns, woods, and coal cabins of the neighborhood, until I was arrested in connection with Richard Laforce for stealing honey, and confined in Poughkeepsie jail. In justice to Richard Laforce I will take the first opportunity to state that he told the truth in his testimony, and that he was not concerned with me in taking the honey. After my discharge from Poughkeepsie jail I returned to Shanandoah.

"On Monday morning, the 9th of October, 1843, I took Mr. Knapp's gun, dog, and ammunition, with five or six balls from the same mould produced on my trial, which I had before secured, and went into the woods with the intention of shooting partridges. When I left Mr. Knapp's I did not think of Mr. Wanzer, nor had any intention of going there. I strolled through the woods on that day, until I reached the Cold Spring Turnpike, passing by Henry Concklin's on my way down, but they did not observe me. I shot at the stump I showed Esqr. Davenport with both barrels of my gun on Monday. I followed the

Turnpike until I reached Thomas Jaycox's. My thoughts at this time were very singular, and I suppose to many incredible, a partial description of which I will give in another place. I went to Benjamin Foreman's barn a little after dark, and slept there till, as I should judge, about eleven o'clock at night. There was something laying heavily on my mind. I wanted to do something, I could not tell what. I almost unconsciously left the barn, took a road leading to Isaac Jaycox's, and thence the road leading to Mr. Wanzer's. I went to his door and made a noise —took hold of the string and raised the latch. Mrs. Wanzer asked who was there? I answered a friend. She inquired what a friend wanted that time of night? I answered to stay all night. I walked away from the door, and laid my hat under a peach tree, about two rods distant. I stood there about five minutes, with my gun cocked and pointed towards the door, intending to shoot him if he opened it. He did not make his appearance, and I retired to his barn and slept there till sunrise the next morning, when I went into the bushes and continued firing my gun at intervals in the neighborhood of his house all the day on Tuesday; once I shot at Mr. Wanzer's fowls. He was at work in his garden and buckwheat the most of the day. At one time I lay within thirty yards of him, with my gun pointed towards him, and said to myself, 'how I will pop him over tonight.' The family all went away at one time, and I took a circuitous route, thinking to go in the house, but on reflection the thought occurred to me, that they might return and find me there. I indulged the hope that Wanzer would come in the bushes, and I would shoot him there. As soon as it was dark I went to his barn and thence to his dirt cellar, and then stood with my gun ready, thinking he would come out. He did come around the corner of his house but went in again immediately. I then went in front of the house, took off my hat and laid it under the same peach tree where I laid it the night before. I whistled to induce Wanzer to come to the door, but he did not come. I went up to the house and looked in at the window adjoining the road. As I looked in, some of the family said, 'hark.' Mr. Wanzer's gun stood up against the wall, he took it in his hand and went to the door. I stood ready to shoot him if he came to the corner of the house. I trembled very much all the time I was there. From thence I went by the dirt cellar into the road and put my gun through the fence. I stepped into the middle of the road, got a stone and threw it against the house. Within a minute after I saw Mr. Wanzer coming down the path with his gun in his arms. He came within a rod and a half of where I lay. My feelings were such that I did not take particular aim. I fired, intending to hit him in the breast, he sprung up, threw back his head, gave a loud groan, and fell apparently without bending, wheeling around at the same time. I then ran into the bushes and whistled for my dog eight or nine times—my dog followed, and I went on through the bushes. I did not go through the oat field—I did not make the tracks Esqr. Wilson testified to. After a little time I stopped to think, and O! how bitterly I regretted that I had shot Mr. Wanzer. I said aloud, 'How the devil will tempt any one, but he shall never tempt me to do the like again; I will get my living hereafter by honest industry or die!' I felt exceedingly dizzy, and did not know what to do. I loaded my gun, but could not recollect immediately after how it was charged. I then drew off the charge and reloaded it, to be certain it would go. I started and ran again, with such feelings as I cannot describe, until I came to a brook, which was running very rapid. Again I stopped to reflect—putting my hand on my head, I discovered my hat was gone. I involuntarily cried out, 'What have I done!' I made a struggle to collect my thoughts—at length it occurred to me that my hat was under the peach tree in front of the house. I thought it would betray me. I went up to the road Ino, my dog, ran up in the woods and barked; he was on the track of something. I said, 'That dog means to betray me yet.' When he returned I drew up my gun to shoot him, when the thought struck me that I should be heard. I mashed him down with my hand, and sat down and listened. I pulled of my boots—left them in the road—took a circuitous route, and got my hat from under the peach tree. No one was at the house—the place was awfully still and solemn. I went on the road towards Isaac Jaycox's some distance, turned up in the woods on a side hill, and laid down all of an hour and a half. While laying there my dog went away. I had not proceeded far from that spot, when on feeling my pocket, I discovered that my powder-horn was gone. I went back to search for the powder-horn, but could not find the place where I had lain. Whilst searching for the horn, my dog met me. Returning to the place where I left my boots, and putting them on, I followed the road towards Isaac Jaycox's a quarter of a mile, when I struck off in a southerly direction, expecting to come out in Isaac Jaycox's open fields, but came out by Samuel Denny's, the husband of Mary Denny, on the turnpike. Then I knew where I was. I passed by Samuel Denny's, but it was all of two hours after I shot Mr. Wanzer. I made an attempt to get in Joseph Ferris' barn, but failed. Leaving Mr. Ferris' barn, I followed the turnpike to Thomas Jaycox's—took his buffalo skin, blanket, and whip from his barn, and lay in the woods till the sun was three-quarters of an hour high the next morning. Then taking a southerly course, I had not proceeded far when I was induced to stop for fear of wetting my gun, as the leaves were very damp. Shortly after I heard a rustling in the bushes behind me, and supposing it might be some one in pursuit, I proceeded on. I lay in sight of Andrew Miller's house some time. Whilst there, Daniel Ferris' boy rode up to Miller's on a horse, and told Mrs. Miller that Uncle Wanzer was shot the night before, and that they suspicioned little George Denny. My object in laying there was to get an opportunity to get something to eat, and to get what money I could find. What Daniel Ferris' boy said frightened me. I then went south, below John Brower's, and crossed the road—went into Joseph Ferris' woods, and turned my course for Shanandoah. As I was going up I heard a wagon coming, and discovered that Andrew Miller and Rufus Gillet were drawing cordwood, but by stepping a little one side, they did not observe me. A little further on, I got some apples off a tree—went across Forge Hill, and came into the turnpike a little west of Henry Concklin's, and following the turnpike till I came to a cross road, which I took. I came out by Elijah Horton's. I talked with Elijah Horton. He took my gun and discharged it. I had conversation with Benjamin Mulcox—he told me that Wanzer was shot the night before in his left side. He intimated that I was suspicioned. A short distance from Mulcox, Wixon and Booth arrested me. They said George, is this you? Booth looked at my gun to see if it was loaded. I told him it was not. They slapped their hands on my shoulders and said, 'You are our prisoner.' I asked them what they meant. Booth said, you are a good-for-nothing murderer. I answered that I had not shot any man. They then took me to Wixon's, and thence to Wanzer's. As to my motive in going to Wanzer's, I do not know what to say. When I came to Thomas Jaycox's on Monday, many thoughts crossed my mind. At one time I thought of going to John Brower's, about three miles south-east from Wanzer's, to kill him and his family—to rob the house, and set it on fire. I thought that Rufus Gillet lived close by, and he would detect me. I also thought of robbing Uncle Joseph Ferris' house, and of killing him and his family; but I said some one lived with him, and I might be detected. I turned from these thoughts. I also thought at one time I would kill Wanzer, his wife, and children—drag them in the house, and set it on fire; but I have doubts whether I was sincere in my reflections to their fullest extent, as, after I had killed Wanzer, I thought no more of his wife or his children. As to the witnesses who testified against me, I have no other desire than to corroborate them all, so far as they have stated correctly, or as far as their testimony related to me. Peter Vantasel, who was the most suspicioned, as near as I can recollect, told the truth. I do not recollect of telling Derbyshire that 'I wanted to shoot three or four, and if Uncle Wanzer did not hush up about the key, he would be the first one.' I told Derbyshire when I bought his gun, that the money I paid him I took from Andrew Miller's trunk, and he promised not to divulge it. I have no recollection of saying to John A. Miller, 'I would fix Uncle Wanzer yet.' John Garrison's testimony had no reference to me; he did not see me. I was but a short distance from Wanzer's house all the day on Tuesday. There were but two kinds of shot in my gun, beside the ball, when I shot Wanzer. I picked the smaller size out when I charged it. And I now solemnly assert, that I had no bitter feelings against Mr. Wanzer at that time; and possess none now against any individual. I do not fear death, but I cannot say how much my mind may change as the hour approaches, and I still have a desire to live as long as I can."

"Here closed his statement, the narration of which in his simple, but I am satisfied candid manner, occupied the space of three hours. Discovering probably that I was anxious to leave, after he had concluded his confession, he begged of me not to be impatient, but to remain with him a little longer. I assented. He read the 2d chapter of James, and requested me to explain the 13th verse. I replied that there were several clergymen residing in and near the village, and if he desired it, I would give them a general invitation to visit him, with which he appeared satisfied. Here I took my leave, not, however, without promising, at his solicitation, to visit him frequently during the short period allotted for his existence.

"Carmel, June 4, 1844."

REVOLUTIONARY ANECDOTES, REMINISCENCES, ETC.

During the year 1779 and '80, Washington frequently crossed the Hudson from West Point, inspecting its outposts, and visiting the eastern States. Daniel Haight, deceased, kept a tavern in the old house, now occupied by one of his sons, on the cross-road, north of Judge Garrison's mansion, leading to the Peekskill and Cold Spring turnpike, at Henry Croft's house. The Commander-in-Chief was in the habit of stopping at "Haight's tavern," to rest himself and suite, in passing to and from Continental Village and the eastern States. It is well known that he was of a contemplative turn of mind, of few words, and not given to "*much speaking.*" Mr. Haight, in speaking of him, long after the war had closed, and long, too, after the "Father of his country" had been "gathered to his fathers," often remarked about the silent, meditative mood evinced by him while at his house. He said he never knew Washington to commence a conversation unless spoken to, or he desired something to be brought to him. He called at Haight's house one day in the fall of 1780, and as he entered the house, the servant girl ran up the stairs, and when half way up, fell; Washington broke into a hearty laugh, and turning round, said to the *host,* "*It is the first time I ever saw a person fall up stairs.*" After he had departed, Haight remarked to his family that it was "the first time that he had ever seen the Commander-in-Chief laugh;" and since that time, he has said, "that it was the last."

James Croft, the father of Henry and Stephen Croft, Esqs., was an enlisted soldier in the Revolution. He was attached to the northern army, and after the surrender of Burgoyne, was discharged. While returning to his home, near, or at Continental Village, he stopped on the west side of the Hudson, near Kingston, where some British officers where enlisting men for their army. They wanted Croft to enlist, but he promptly refused. On pretence of shaking hands with him as he was about to depart, one of the officers dropped a gold piece in his hand, and then said to him that he had had the King's money, and "*he'd be damned* if he shouldn't enlist," or pay a fine. Croft steadily refused to do either, and told the officer that when he enlisted again, it would "be in the cause of his country and her rights." He had $18 in hard money with him, it being the balance due him when his term of enlistment expired.

The British officers forthwith, organized a drumhead court-martial, fined him to that amount, and took it by force from him. Thus penniless, he returned to his home. Reader! the incident is a small one, but should you be ever similarly situated, stand fast to your integrity and your country ; and remember this little incident that records the patriotism of James Croft.

When the unfortunate Major Andre was arrested by Pauldings, Williams, and Van Wart, they first conveyed him to North Castle, in Westchester County, where Col. Jamieson commanded. Afterwards he was conducted by a guard of twenty men, to Salem, the quarters of Col. Shelden. During their journey thither, they stopped for the night at Comyen hill, where Major Tallmadge, the commander of the guard, tied Andre to a tree, as an additional security. From Salem, he was brought to the Red Mills in this county, and lodged for the night at the house of James Cox, a Major in the Ordnance Department of the Army, and grandfather to the wife of the Hon. John Garrison, of Philipstown. While here, two soldiers were stationed at the door, and two at each window of his apartment. Phœbe Cox, a daughter of the above-named Major, and the mother of Mrs. John Garrison, was then an infant, laying in the cradle when Andre entered the room. Andre stepped to the cradle, and the child, which had just awoke, looking up at him, smiled. His feelings seemed immediately touched, and, in a tone of deep melancholy tenderness, he said, "*Happy childhood! we know its peace but once. I wish I was as innocent as you.*" From the Red Mills, he was brought by the way of Continental Village to the "Robinson House" in this town, under a guard of one hundred horse, by Major Tallmadge. While at the Red Mills, and looking in a mirror in his room, he saw a hole under the armpit of his coat, and perceiving that the officer who was in attendance observed it also, he smiled and remarked, "I presume Gen. Washington will give me another coat." From the "Robinson House," he was conveyed to West Point; from there to Stony Point; and from thence to Tappan, or Orange town.

The notorious Joshua H. Smith, to whose house, two-and-a-half miles from Stony Point, Arnold conducted Andre, after their midnight interview "at the foot of a mountain called the Long Clove, near the low-water mark," was arrested at Fishkill, Dutchess County, and brought to the "Robinson House," a few hours before the arrival of Andre at that place. He had furnished Andre with a coat, saddle, and bridle, and after secreting him all day, conducted him the night after Arnold's interview with him, to the ferry at Stony Point, crossed over to Verplank's Point, and slept with him at a house near Crom Pond. The next morning they started as soon as it was light, and rode as far as Pine's bridge, where they halted and made a breakfast of "*suppon*" and milk. Smith here left Andre, and giving him some Continental money, advised him to take the road to White Plains. Six miles beyond this, Andre was arrested. Smith was tried before a court-martial, and imprisoned in the jail at Goshen, Orange County; escaped from thence to New York, and returned with the British army to England, where some years ago, he published a little volume entitled "Major Andre," in which he gives an account of his relations with Arnold and Andre, his arrest, trial, and imprisonment; and endeavors to show that he knew nothing of the real business between the British Adjutant-General and America's great traitor, with an outpouring of abuse on Washington, Green, and other patriots, that would disgust any man but an Englishman, and an American tory. The book has no date, but was published a few years before his death. We quote from that part of it that relates to his arrest and arrival at the "Robinson House," with his interview with Washington.

"Having given him (Andre) directions about the road he was to take upon crossing the bridge, with a message to my brother, the chief justice, whom he knew, we parted. I proceeded on my way to Fishkill, taking Gen. Arnold's quarters at Robinson's house in my route: I mentioned to Gen. Arnold the distance I accompanied Mr. Anderson, which gave him apparently much satisfaction. His dinner being ready I partook of it, refreshed my horses, and in the evening procceded to Fish Kill to my family. Here I found General Washington had arrived in the course of the afternoon, on his return from visiting Count Rochambeau, and I supped in his company with a large retinue, at Gen. Scott's. The next day I went on business to Poughkeepsie, and returned to Fish Kill the ensuing evening. It was on the 25th of September, about midnight, that the door of the room wherein I lay in bed with Mrs. Smith, was forced open with great violence, and instantly the chamber was filled with soldiers, who approached the bed with fixed bayonets. I was then, without ceremony, drawn out of bed by a French Officer, named Govion, whom I recollected to have entertained at my house not long before, in the suite of the Marquis de la Fayette. He commanded me instantly to dress myself, and to accompany him to General Washington, having an order from the General, he said, to arrest me. The house was the residence of Col. Hay, who had married my sister. The family was thrown in great confusion; the female part especially were in the deepest distress; indeed the shock so much affected Mrs. Smith, that she never fully recovered from it; and, which added to my subsequent sufferings, was the cause of her death. I perceived that any opposition would be ineffectual. Col. Hay desired to know for what cause the arrest was made; to which Govion would give no satisfactory answer. I then desired the privilege of having my servant and one of my horses to go with him to Gen. Washington, at Robinson's house, which he refused; and I was immediately marched off, on foot, the distance of 18 miles.

" At length on my arrival at Robinson's house, I was paraded before the front door, under a guard. General Washington soon afterwards came into a piazza, and looked sternly and with much indignation at me; my countenance was the index of my mind, and the beautiful lines of Horace occurred to me, '*si fractis et illabiter orbis inupavidum feriunt, que ruinae*,' &c.

" On his retiring, I was ordered into a back room, and two centinels placed at the door.

" After as much time had elapsed as I supposed was thought necessary to give me rest from my march, I was conducted into a room, where were standing Gen. Washington in the centre, and on each side Gen. Knox and the Marquis de la Fayette, with Washington's two aids-de-camp, Colonels Harrison and Hamilton.

" Provoked at the usage I received, I addressed Gen. Washington, and demanded to know for what cause I was brought before him in so ignominious a manner?—The General answered sternly, that I stood before him charged with the blackest treason against the citizens of the United States; and that he was authorized, from the evidence in his possession, and from the authority vested in him by Congress, to hang me immediately as a traitor, and that nothing could save me but a candid confession who in the army, or among the citizens at large, were my accomplices in the horrid and nefarious designs I had meditated, for the last ten days past.

" I answered that no part of my conduct could justify the charge, as Gen. Arnold, if present, would prove; that what I had done of a public nature was by the direction of that General, and, if wrong, he was amenable; not me, for acting agreeably to his orders.

" He immediately replied, 'Sir, do you know that Arnold has fled, and that Mr. Anderson, whom you have piloted through our lines, proves to be Major John Andre, the Adjutant-General of the British army, now our prisoner? I expect him here, under a guard of 100 horse, to meet his fate as a spy, and, unless you confess who were your accomplices, I shall suspend you both on yonder tree,' pointing to a tree before the door. He then ordered the guards to take me away.

"In a short time I was remanded into the room and urged to a confession of accomplices, with Gen. Washington's declaration, that the evidence he possessed of my being a party, was sufficient to take away my life.

"The Gen. irritated by my reply, remanded me back to my confinement.

"Some time afterwards, Col. Hamilton came to me, and compassionately, as he said, recommended me to declare all I knew respecting the business of which I was accused, observing that many were mistrusted, who, if they confessed, would be in a worse situation; but as he supposed this was not the case I had now a chance to save my life, and for the sake of my family I ought to preserve it,—with many more expressions to the same effect, &c.

"Gen. Washington then came into the room, and questioned Col. Hamilton why he was so long speaking to me? The Col. replied, 'General, I know Smith has meant well during his agency in this transaction, for in all our public meetings in New York his general demeanor spoke a spirit of moderation, nor could he be persuaded to any other opinion than that this contest between Great Britain and her colonies would be compromised, as in the business of the stamp and other acts of which we comcomplained to the British Government, in our petition by Gov. Penn,' &c.

"Gen. Washington then said in a gentle tone of voice, 'Col. Hamilton, I am not yet satisfied; take him into the back room; we must know something more about this business.' I was then conducted into the recess from whence I was brought.

"I was about to take some refreshment, when one of the sentinels, posted at the door, vowed that if I touched any of the biscuits that were in the room, he would shoot me dead.

"The fact was that the room was a kind of a butlery, in which Mrs. Arnold had placed her stores, and I was in the act of taking a piece of the biscuits. I made no reply to the sentinel; but remained nearly two hours in this confinement, when I heard the tramp of a number of horses near the place where I was confined, and, soon after, could distinguish the voice of the unfortunate Andre, and of Gen. Washington and his suite, who soothed him with all the blandishments that his education and distinguished rank demanded; he was courted with a smile in the face, when worse than a dagger was intended for his heart. I distinctly heard Col. Hamilton say to a brother officer, who came out of the same room, that Major Andre was really an accomplished young man, and he was sorry for him, for the Gen. was determined to hang him.

"It was nearly dark, when a very respectable young gentleman entered the room, and politely desired me to accompany him. I was in hopes this was a prelude to my emancipation, and I requested the honor of his name? He answered, 'It is Washington.' I said, 'I presume, Sir, you hold the rank of Colonel?' He told me he held no rank at all. He then conducted me to the back part of Robinson's house, where there were two horses, desired me to mount one of them, and by his guidance in a way I had never been, we soon reached the bank of the river opposite to West Point. Here I was delivered to the custody of a Capt. Sheppard, of the New Jersey Continental troops, and did not observe I had been guarded by a troop of horse until I was placed in the ferry-boat, and saw them follow Mr. Washington up the mountain; two boats followed us, composed of the guard. If I had any inclination to throw myself overboard, I was so well guarded, that I am certain I should have been taken out of the water; for the main object of Gen. Washington in detaining and trying me, was to obtain a knowledge of Gen. Arnold's confederates in the army, as well as in Congress. In fact, this defection of Arnold had excited such a general suspicion, that no one dared trust another; and nothing but execrations were heard from hut to hut."

John Warren, Esq.—This gentleman was of English origin, and the son of Samuel Warren, who came from Boston to this town previous to the Revolution. John, the subject of this sketch, was born in 1765, and had five sons and two daughters, viz.: Cornelius, Sylvenus, Samuel, Harry, and John, who died October 8th, 1840, Mary and Susan, who are both dead. Mary married Joseph Haight; Susan married Elijah Davenport. He was born in the house now occupied by John Hopper, Esq., on the hill south of the Warren Mill, now owned by Justus Nelson, on the old post road. His father, Samuel Warren, resided in it during the Revolution, and the place went by the name of Nelson's Highlands. Samuel, the father of John, was killed at the Franklindale bridge, by being thrown out of his wagon. Some of the planks were off, and the horses taking a turn, he was thrown from the wagon, through the aperture into the stream below.

John Warren was married when 17 years old, and his wife 15; and, for some years thereafter, lived in a log-house which stood in the corner of the orchard, across the road and opposite to the residence of Justus Nelson, Esq., which he also built with the grist mill attached to the same property. This farm, consisting of 300 acres belonged to him. He was a blacksmith by trade, and served his time with Peter Warren, father of James Warren, Esq., of Cold Spring, who lived where Ja. Griffin, Esq., at present resides in the lower part of this town. The first knives and forks he had when he commenced keeping house he made in his blacksmith shop. Shortly after he was married, being in want of a pair of pantaloons, he went to Peekskill and asked a merchant to trust him for the cloth; the merchant refused, and Warren returned home without making further application for credit. He told his friends that he had made up his mind, "that the best policy for a man was to pay as he went." This was what the celebrated John Randolph called the "*philosopher's stone.*" In the midst of a stormy debate in the House of Representatives, Randolph arose and screamed out at the top of his voice, "Mr. Speaker, I have found the 'philosopher's stone'"—it's pay as you go." This was the first and the last time that John Warren asked for credit; and in a few years his industry and prudent habits placed him in a condition in which everybody would have been glad of an opportunity to trust him. During the whole course of his life he never sued nor was sued by any person. He began poor and died rich; free from debt; and what he had acquired, he obtained in and by virtue of that curse, originally pronounced upon all mankind through Adam, as our federal head and representative, "by the sweat of his brow."

He aspired to no higher distinction than that of a plain, practical farmer, which he was. The purity of his motives, and the honesty of his heart, were never questioned; and in all the relations of life he never gave just cause of offence to his neighbor. He died, regretted and beloved by all who knew him, in 1837, in the 72d year of his age. His children, so far as we know them, inherit his virtues.

Capt. Samuel Jefferds.—This gentleman was born in Boston, Massachusetts, in 1752, and entered the Continental army at the age of twenty-three, as first Sergeant in a company of artillery. He served through the war, and for three years after peace was concluded. He was stationed two years at Fort Stanwix, erected in the year 1758 by the English, and named from Gen. Stanwix. "It occupied a position commanding the carrying place between the navigable waters of the Mohawk and Wood Creek, and was regarded as the key to the communication between Canada and the settlements on the Mohawk." During the Revolution it was named Fort Schuyler. He was afterwards stationed at Fort Putnam, West Point, and its out-posts. He fought most gallantly in the battles of Bunkerhill, Long Island, Brandywine, and Monmouth. The battle of Monmouth was fought on Sunday, and the day was intensely hot. A great many soldiers died from over-exertion, the heat, and drinking too much cold water.

In speaking of the engagements he had passed through, after his retirement from the army, to his only surviving son, Samuel Jefferds, Esq., of Philipstown, he remarked, that, at the battle of Monmouth, "the heat was so intense that it nearly exhausted his nature."

Four years after the war he married the widow of Peter Warren, who was the father of James Warren, Esq., now of Cold Spring. After his retirement from the army, he lived where Ja. Griffin, Esq., now resides on the old post road in the south part of this town, and a short distance north of Continental Village. He was quartered here in the winter of 1779, '80.

He was plain in his manners, of a kind and gentle disposition, but when provoked from a sense of injury, energetic and strongly moved. He had battled long for the liberties of his country, and doubly appreciated the blessing of peace when it came. Distinguished for strict integrity and frankness of disposition, he had no enemies while living, and no calumniators when dead. Amid the carnage, the smoke, and the thunder of battle, his courage never forsook him. It was of that cool, deliberate character—not created by the occasion and the excitement of circumstances—which has its foundation in a constitutional disposition. Amid the roar of cannon and the conflict of arms, he would ride round and direct the operations of his batteries with the calm steadiness of feeling, that would characterize others only at a review. He passed through a seven years' war, and four hard fought and well contested actions, unscathed, enjoying the confidence of Washington, Knox, Green, Hamilton, and their associates; and was elected a member of the Society of the Cincinnati, in 1785. This fact alone shows the estimate which Washington and his compatriots, placed on his patriotism, services, and character. It was not every officer of the rank of captain, who could get admission into this honorable Society of banded brothers.

The following is a copy of his certificate of membership, now in the possession of his son, Samuel Jefferds, Esq.:

"Be it known that Samuel Jefferds, Esquire, Captain of Artillery in the late Army is a Member of the Society of the Cincinnati instituted by the Officers of the American Army, at the Period of its Dissolution, as well to commemorate the great Event which gave Independence to North America, as for the laudable Purpose of inculcating the Duty of laying down in Peace Arms assumed for public Defence, and of uniting in Acts of brotherly

Affection, and Bonds of perpetual Friendship, the Members constituting the same.

"In Testimony whereof I, the President of the said Society, have hereunto set my Hand at Mount Vernon in the State of Virginia, this tenth Day of December, in the Year of our Lord One Thousand Seven Hundred and Eighty Five, and in the Tenth Year of the Independence of the United States.

"By order,

"G. WASHINGTON, President.

"KNOX, Secretary."

The badge of the Society is a bald eagle suspended by a blue ribbon, edged with white, emblematic of the union of France and America. On the breast of the eagle, Cincinnatus is receiving the military ensigns from the three senators; the implements of husbandry are seen in the back-ground; round the whole, *Omnia reliquit servare rempublicam*,—'He relinquished everything to serve the country.' On the reverse, Fame is crowning Cincinnatus with a wreath, inscribed, "Virtutis præmium,"—"*the reward of virtue*,"—with other emblems; round the whole, "Societas Cincinnatorum, instituta A. A. 1783."

This society was formed by the surviving officers of the Revolutionary Army, to perpetuate their friendship, and to raise a fund for relieving the widows and orphans of brother soldiers who were killed during the war. The idea of establishing it, it is said, originated with Gen. Knox, who, as his biographer says, was the first Vice-President, and continued such to the day of his death. If this be so, he was also Secretary at the same time. The original certificate, now lying before me, given to Capt. Samuel Jefferds, is in the handwriting, and signed by Knox, as Secretary. Washington was the first President, and held the office till his death.

The Society was named in honor of Lucius Quinctius Cincinnatus, a patrician of the Roman Republic, who, while at his plough, was twice called to deliver his country from her enemies; and having done so, resigned the supreme power which had been entrusted to him, he returned to his little farm to continue his rural labors.

Some republicans, at that early day, opposed the society, alleging that it contained the elements of future aristocracy; and among the number was the philosophic Franklin. There are said to be seven State Societies, which hold a general meeting once in three years. By its constitution, the honors of this association were to be "hereditary in the eldest male line of the original members, and, in default of male issue, in the collateral male line."

Capt. Jefferds died in August, 1804, aged 52 years, leaving one son, an only child, Samuel Jefferds, Esq., of this town; who inherits the honors of the Society of Cincinnati, by virtue of the original membership of his father.

The West Point Foundry.—This institution, one of the largest of its kind in this country, is situated about half a mile south-east of the steam-boat landing, in Cold Spring. During the last war with England, some difficulty was met with by the General Government in procuring a sufficiency of large ordnance for the use of the army, as but one foundry, we believe, at that time, was in operation where the largest class of guns were cast. We have been informed that, in view of this fact, this institution was created and set in operation as a matter of pride, by a dozen or more individuals, who received assurances from the Government, that a certain proportion of *Government work* should be done in it, if the establishment was not located nearer than fifty miles to the city of New York.

In the spring or summer of 1817, the Association purchased 150 acres of land at Cold Spring, of Capt. Frederick Philips, and erected a moulding-house, boring mill, a blacksmith shop for light work, pattern shop, a drafting office, with a general office for the use of the Association.

In the winter of 1818, the Hon. Gouverneur Kemble, Joseph G. Swift, James Renwick, and others, petitioned the legislature for an Act of incorporation; and, on the 15th of April, in the same year, an Act was passed, constituting "Gouverneur Kemble, James Renwick, Henry Brevoort, Jr., Joseph G. Swift, John R. Fenwick, William Kemble, Henry Cary, Charles G. Smedburg, Nicholas Gouverneur, Robert J. Renwick, and William Young, and such others as might thereafter, be associated with them, 'a body corporate, in fact and in name, by the name of *the West Point Foundry Association.'*"

Section 3rd of the Act "further enacted, That Gouverneur Kemble, James Renwick, Henry Brevoort, Jr., Wm. Kemble, and Charles G. Smedburgh, shall be the first directors; and the said Gouverneur Kemble shall be the first president of the said company," who were to hold their offices until the first Monday of May, in the year 1819. In 1839 the finishing or machine, smith's, and boiler branches of the establishment, whose operations were carried on in the city of New York previous thereto, were transferred to Cold Spring, where large and commodious buildings were erected. It employs from four to six hundred men, including a few boys; and has a foreman at the head of each branch. Its plan, with a description of its buildings, their size and number, are as follows, viz.:

1. Moulding House, 218 feet long, 68 feet wide, with brick walls 16 feet thick. Roof of slate. This building has 2 cupola furnaces.

2. Gun Foundry, 95 feet long, 75 feet wide, with stone walls, 29 inches thick, and shingle roofed. This building has 3 air furnaces.

3. The Boring Mill is in two buildings; the first 57 by 63 feet; the second 47 by 54 feet. This mill is driven by a 36 feet over-shot wheel, and has an auxiliary steam engine. Its walls are stone, 24 inches thick; roof of tin. The boiler of the steam engine is placed outside of the building, in the yard.

4. The Blacksmith Shop is 128 feet long, 54 feet wide; walls of brick, 24 inches thick; roof of tin. This shop is driven by a water wheel under an 105 feet fall of water, but has an auxiliary steam engine. The engine and boiler are in a separate shop at the east end of the building.

5. Small Blacksmith Shop, 40 feet long, 30 feet wide, with brick walls, 18 inches thick; and a tin roof.

6. Blacksmith Shop for small forges, 52 feet long, 25 feet wide; frame building; and roof of wood. These forges used occasionally.

7. Turning Shop, 60 feet long, 37 feet wide, with brick walls and roof of tin.

8. Finishing Shop, 251 feet long by 54 feet wide, brick walls, and tin roof. The Pattern Shop is on the second floor of this building.

9. W. P. F. Office, 42 feet long by 23 feet wide; a frame building with shingle roof. The Drafting Office is in the second story.

10. Boiler Shop, 100 feet long by 45 feet wide; frame building with shingle roof.

11. Punching Machine House, is 22 feet square; frame building with shingle roof.

12. Coal House.—A stone building with shingle roof.

13. Store, 14 feet by 32 feet; brick walls and shingle roof. Now used by finishing department.

14. First Pattern House, 50 feet long, 36 feet wide; frame building with shingle roof.

15. Second Pattern House, 60 feet long, 30 feet wide; frame building with shingle roof.

16. Third Pattern House, 50 feet long, 36 feet wide; frame building with shingle roof.

17. Fourth Pattern House, 88 feet long, 30 feet wide; frame building with shingle roof.

18. Fifth Pattern House, 40 feet long, 30 feet wide; frame building with shingle roof.

19. Fire Engine House, 19 feet long, $8\frac{1}{2}$ feet wide; frame building with shingle roof.

In addition, the Association also own a large number of houses and building lots in the village, and on the high grounds adjacent to the Foundry.

The site it now occupies was formerly a swale, at the mouth of what was called, in old deeds, the "Margaret Falls' brook." This brook discharged itself into the shallow bay or *flat*, which lies between Constitution Island and Cold Spring. The foundation wall of the old Moulding House, on its south side, was sunk down sixteen feet to the solid rock; while the north wall, resting on the same rock, was sunk only six feet. The dip of the rock is about forty-five degrees.

The Blast Furnace, just north of the Blacksmith shop, is not now in operation. Up to June, 1844, large quantities of iron ore were smelted here, obtained from the mines in the Highlands of this town. The cost of quarrying, transportation, blasting, &c., was found to be greater than purchasing it abroad, in a state fit for immediate use.

In the Blacksmith shop there is one trip-hammer of eight tons weight, and two tilt-hammers,—one of 1000 and the other of 500 lbs. Shafts of two feet diameter have been forged here weighing fifteen tons.

The Machine Shop contains thirteen turning lathes and one drilling machine.

The Pattern Shop has four turning lathes, a whip-saw, and planing machine.

In the Boring Mill are eight gun-beds for boring cannon, one slot machine, four turning lathes, two planing machines, three drilling machines, and one large bed for boring cylinders.

The Finishing Shop contains four vertical and one horizontal drilling machines, and four planing machines.

In the Boiler Shop are three punching machines, one riveting machine, and one shearing machine.

The new Boring Mill, now erecting, will contain

one machine for slotting, planing, and drilling, with one large-faced lathe.

The consumption of the principal materials during the year, from March, 1847, to May 1st, 1848, was as follows :—Pig iron, $197,434; coal, $38,405; bar iron, $62,562; boiler iron plate, $37,988; copper, $17,392; total, $353,781.

The principal articles manufactured during that year were, 20 steam engines; 90 32lb. guns; 4 12lb. guns; 2 9lb. guns; 4110 tons of pipes for the Boston Water Works; 1040 tons of large wrought iron work; together with a large number of rolls, wheels, steam boilers, and machinery of various kinds.

The Hon. Gouverneur Kemble, has been the President of this Institution from its incorporation until the expiration of its charter in 1843 or '44, we believe, with the exception of four years while representing this district in Congress: during which, that office was filled by his brother, William Kemble, Esq.

The Vice President, for several years previous to that time, was the Hon. Judge Parrott, late a Captain in the Ordnance Department of the U. S. Army, who still fills the office.

The Association is now conducted by Mr. Kemble, as lessee.

Under the supervision of these gentlemen, everything moves on with the regularity of clock-work. The men receive their wages every two weeks; and work but ten hours per day. If the necessity of the work requires them to work longer, they are paid accordingly. In the blacksmith shop, the fire is not out of the furnaces for weeks; one *gang* of men working through the night until morning, when their places are supplied by another gang during the day. The means and appliances sometimes attempted to be used by other manufacturing establishments to control the political sentiments of their workmen, are not countenanced here. From the most finished workman, down to the smallest boy who twists a rope of straw, there is the most perfect freedom of thought and action in everything appertaining to a man's religious and political faith.

This institution is the life of Cold Spring Village and Nelsonville, and, with "more truth than fiction," it may be said, it feeds all, clothes all, and supports all.

Were it discontinued, those villages would soon look like an *infected district*—resemble, somewhat the city of New York on the first day of May, when everybody, apparently, is *moving*—and realize "Goldsmith's description of a deserted village."

Mr. Kemble has so long superintended this institution—watched it in its career of infancy, until it has attained its present growth, that it may well be called "the child of his affections;" from whom it were now, well-nigh, socially impossible to part.

A branch railroad is now constructing from this establishment, to intersect the Hudson River Railroad, which crosses Constitution Island and the *bay* or *flat*, a short distance west of it. A dock also is being built from the present Foundry-dock, extending west to the channel of the Hudson.

The real and personal estate of this Foundry, in 1846, was assessed at $88,000. It probably now amounts to $150,000.

In 1844, the United States Steam Revenue Cutter, Spencer, of iron, was built here. During the same year another iron steam vessel called the Margaret Kemble, to ply on one of the southern rivers, was also constructed at this establishment.

"Machinery of the most delicate and complicated arrangement, and the engines of the fastest steamboats in the world, are manufactured here.

"The following circumstance is no mean compliment to the mechanics of the New World, and a plain demonstration that 'Yankee ingenuity' is known and appreciated even beyond the waves of the Mediterranean: The Pacha of Egypt was anxious to obtain a machine, among several others, which would take the hull from cotton seed. Having applied to several mechanics of Europe without success, he resolved to try what American invention could effect; no doubt very wisely thinking, that a people who could manufacture wooden hams 'almost as good as the genuine article,' and baswood cucumber seeds 'so perfect that they will vegetate,' could most certainly furnish any kind of machinery which he might want. The application was made, and in a comparatively short time a machine was finished at the West Point Foundry, which, at least in the opinion of the inventor, would answer the purpose of the Pacha. It was shipped from New York in January, 1838, accompanied by an experienced engineer, who was sanguine in the belief that the experiment would be successful."

TOWN OF PUTNAM VALLEY.

This town lies wholly in the Highlands; and, like the greater part of Philipstown, from which it was taken, is rough and mountainous. Iron ore is found here in abundance, but its distance from the river, and the absence of easy facilities, will prevent, for some time to come, attempts to unlock its mountain repositories. The valleys, Canopus and Peekskill Hollows, are rich, fertile, and well cultivated. They stretch, like a pair of garters, through the entire length of the town from north to south. It is centrally distant about twelve miles west of Carmel. Its population in 1840 was 1,659: and in 1845, 1,598. Having treated of its early settlement under the head of Philipstown, and delineated its geological features in the article *Geology*, there remains but little to be added.

Oregon.—A small village, about three miles east of Annsville, near the Westchester line. The Peekskill Hollow Creek and the out-let of the Horton Pond meet at this village, and form one stream.

Crofts.—A few houses in Canopus Hollow. A store and tavern were kept here; the latter has been discontinued. It was formerly called Sodom.

Tompkins's Corners.—A few houses at the intersection of the old Wickopee and Peekskill Hollow road.

Hempstead Huts.—These buildings are revolutionary relics, and are located on the farm of Harry Gillet. A detachment of the troops of the Massachusetts Line, with a company or two from Hempstead, Long Island, occupied them in the winter of 1780; which accounts for the name. The chimneys still remain, but the huts have been burnt down.

Canopus Hill.—An eminence in the south-west part of the town, on the farm of —— Meeks, Esq. It is named in honor of an Indian Chief.

Tinker Hill.—This hill is about three miles northeast of Canopus Hill, and is owned, partly, by John Odle, Esq. About fifty years ago an old Englishman lived on it, named Cornelius Rick, who went about *tinkering*, or doing a little at every trade; and hence the name.

Ponds.—There are nine ponds in this town, some of them of more than ordinary size. The largest is the

Horton Pond.—It is located in the centre of the town, and contains excellent Bass and Pickerel, some of the latter weighing six and seven pounds. It is bounded by the lands of Lee Horton, Abijah Lee, Charity Smith, Wesley Christian, Solomon Baxter, Daniel Barger, Joseph Strang, and Henry Mead; and is about $1\frac{3}{4}$ of a mile in length and 1 in breadth. It is named after the Horton family who formerly owned all the land adjacent to it.

Solpeu Pond.—This pond lies in the west part of the town, about one mile from the Horton Pond; and was named after a person of that name who lived in its vicinity. It is about a mile in length, and half a mile in breadth. Like the Horton Pond, it contains excellent Bass and Pickerel. It is circumscribed by the lands of Mrs. Ann Horton, William Jerry, James Likely, and William Denny.

Barger's Pond.—This sheet of water is in the south-east part of the town, about three-fourths of a mile in length, and one-fourth in breadth. It takes its name from the Barger family, who have long resided in the neighborhood of the pond.

Bryant's Pond.—This pond is next in size; and is situated in the east part of the town, about three-fourths of a mile north of Barger's Pond. It is half a mile in length, nearly that distance in breadth, and is named after Solomon Bryant.

Muddy Pond.—This body of water lies about half a mile north of the Horton Pond, is nearly half a mile in length and one-fourth in breadth, and is called from the *muddy* appearance of its water.

Clear Pond.—This beautiful sheet of water does not belie its name; for a more pellucid body of water is not found in Putnam County, or in any other. It is in the north part of the town, has no inlet, is formed by springs, and clear as crystal. It is about half a mile long, nearly that distance in width; runs south into the Muddy Pond, which empties into the Horton Pond, which, in turn, empties into the Creek at Annsville, just south of the Putnam and Westchester line. Those who live near it, and are familiar with its water, assert, that a person can see twenty feet, or more, into it.

Jonathan Owen's Pond.—A handsome sheet of water about half a mile long and one-fourth of a mile broad, in the most southern part of the town, and nearly one-fourth of a mile from the Westchester line. It runs into the creek at Annsville above-mentioned. The name explains itself.

Cranberry Pond.—This is the smallest pond in

this town, and is situated about half a mile north of Owen's Pond. It is about one-fourth of a mile in length and forty rods in breadth; and located on the land formerly owned by Philemon Smith. Immense quantities of *Cranberries* grow upon the low grounds skirting it, and hence the name. *Cran* is saxon; and berry is derived from the Saxon word *beria*. Compounded, it means a berry that grows on a slender, bending stalk; also called *moss-berry*, or moor-berry, as it grows only on peat-bogs or swampy land.

Pelton's Pond.—A small body of water lying north of the Clear Pond, located in the north part of the town, one-fourth of a mile long and about one-eighth in breadth. It takes its name from a man who worked an ore-bed near it, and empties into the stream running to Annsville.

Peekskill Hollow Creek.—A small stream rising from a spring just south of Stillman Boyd's, in the town of Kent, running the entire length of the Hollow, and falling into the outlet of the Horton Pond. On this stream are the following mills, viz.: Herman Adams's saw mill; John Post's saw and grist mill; Pratt's trip-hammer, turning-lathe, and whip-saw works; Thomas Winter's grist mill, and John Sillick's saw mill; and the wire factory belonging to Joseph Strang & Co.

Canopus Hollow Creek.—This stream rises near the second *Gate* on the Cold Spring turnpike, runs through the Sunk Lot, and falls into the stream at Annsville, near Peekskill, Westchester County. On it are Bunnell's Forge and saw mill; a saw mill formerly owned by John Horton, and Mowyat's paper mill.

A man named Robert Oakley, who was a staunch Whig, while his brothers were rank Tories, lived, during the Revolution, just above the residence of Doctor John Tompkins, on the Wickapee road. His brothers gave information to the infamous Cunningham, the provost marshal of the city of New York, concerning his opposition to the British cause, who sent a band of Tories to waylay and shoot him. Oakley had been absent from his house, and returned a little before sunset in the fall of the year previous to the *hard winter*, which, we believe, was in 1780. They concealed themselves near his house, and no sooner had he dismounted, than they shot him.

Thomas Richards, who also lived in this town, and was a turner of wooden dishes by trade, was taken as a rebel, and carried down to the Sugar House in New York. His wife was left at home, and the *hard winter* coming on, the snow covered his lowly cabin, preventing ingress or egress by the door. His wife, having first used up all the fuel inside, with the ax broke a hole through the roof, got out and cut the large limbs down, which hung over the small hut, which she threw down into the garret for present and future use. Her stock of provisions soon became exhausted; and their cow, which had been kept in the work-shop, died. This lone woman, without a human being for her companion, and confined in her prison of snow, was forced to eat the dead body of her cow; and when that was gone, she lived on a little shelled corn that was left in the garret, making use of some filthy, dirty brine, in the bottom of an old pork barrel, to season it with. In this manner she made out to live through the winter. A grandson of

Richards, we are informed, is now living in Fishkill, Dutchess County. Such iron energy and indomitable courage, when called upon to battle with cold, hunger, and thirst, in a dreary solitude, would be found but rarely, if at all, at the present day. The human heart shudders in contemplating the possibility of such a case.

TOWN OF CARMEL.

This town was taken from Frederickstown, at that time embracing the now towns of Kent and Patterson, in 1795; and is centrally distant from New York City about 55 miles, 106 from Albany, 16 east of the Hudson, and 18 from Peekskill. Its soil is a mixture of loam and gravel, with a rolling surface, indented with slopes and vales. It is well adapted to grazing; and large quantities of beef, lambs, sheep, fowls, and other species of "marketing" are produced here for the New York market.

The New York and Harlem Railroad, which is now being extended near its eastern boundary, will greatly facilitate the transportation of its products to market, and enhance the value of the land. It is named after a mountain in Palestine, on the southern frontier of Galilee, constituting a part of Lebanon, in the pachalic of Acca.

From its supposed resemblance to Mount Carmel, "which consists of several rich, woody heights, separated by fertile and habitable valleys," it was christened, at its organization, as above.

EARLY SETTLEMENT.

The first settlement that we have been able to ascertain, was made in this town by George Hughson, who located himself on the ridge just north of Lake Mahopac, and west of the residence of Nathaniel

Crane, Esq., about 1740. A year afterwards William Hill, father of the present William, now living in this town at a very advanced age, and his brother Uriah, came up to the Red Hills, when William, who was the younger of the two brothers, was only 12 years old. Their father was Anthony Hill, who came from Holland about the year 1725; and after remaining a short time in New York City, removed to the Fox Meadows, where he made a purchase and settlement.

On the voyage, the whole family, except himself and two sisters, died. Anthony, at about the age of twenty, married Mary Ward, who also came from Holland, by whom he had five sons and four daughters, Uriah, William, Anthony, Andrew, Cornelius, Charity, Jane, Mary, and Merriam.

Anthony Hill died at the Fox Meadows, and his wife at the Red Mills, aged 93. He having bought a tract of land of the Indians, near the Red Mills, he sent his two oldest sons, Uriah and William, to clear it up. Uriah, in some way or other, became obnoxious to the Indians, and was compelled to go back to Westchester.

William remained, and one night going out to look for the cow, which the brothers had brought from their father's farm at the Fox Meadows, he was attacked by some wolves. By climbing up a tree and remaining on it nearly all night, he escaped from them by a circuit to the north side of Lake Mahopac, where early in the morning he came to the log-house of George Hughson. This was the first he knew of a white man residing there. Hughson told him that he had settled there about a year before. At 25 William Hill married a sister of Abraham Smith, the father of the Hon. Abraham and Saxon Smith of Putnam Valley.

William, son of Anthony, and father of the William now living, had eight sons, viz.: Noah, Solomon, William, Cornelius, Abraham, Andrew, two having died in infancy without a name; and four daughters, viz.: Phoebe, Mary, Chloe, and Jane. Noah lived near the Red Mills, and died there in 1830. Solomon lived at the Nine Partners in Dutchess County.

At this time the first house erected in this town was about one mile south of the Red Mills, occupied by a man named Philips, where Ezekiel Howell resides; the next, north, was William Hill's; and the next, George Hughson's. Soon afterwards, the Cranes, the Berrys, Hedyers, Austins, Roberdeaus, and others, settled down in the vicinity of the Hills and Hughsons.

Jabez Berry settled where Elijah Crane now lives, about one mile north of Lake Mahopac. A family of the name of Shaw soon settled at Carmel village, on the north and south shores of the lake which still bears their name.

A short time after the Hills and Hughsons settled, John Carpenter came from North Castle, now called New Castle Corners, and settled where the Hon. Azor B. Cranes resides.

The Carpenter family were Quakers, of English origin, and came from England to Plymouth; but were driven, by persecution at that early day against the Quakers, to Long Island, from there to North Castle, and from thence it came to this town.

John Carpenter's old house stood at the foot of the hill, just south of the residence of the Hon. Judge

Crane, on the east side of the road. The tories, royalists, and the friends of the King, called him the "*damned old* rebel." He was a patriot of the staunchest kind; and if adherence to the cause of his country and her rights constituted a *rebel*, he was one in every sense of the term and in the widest latitude of the expression.

He left his farm to John Crane, who married his daughter. John left it to his brother Joseph Crane, who devised it to its present owner, Hon. A. B. Crane, Judge of the Putnam County Court.

A family by the name of Hamblin settled in this town about the same time with the Carpenters, in the vicinity of Lake Mahopac.

In 1770, John Crane, father of Nathaniel Crane, Esq., now living, built the first frame house in this part of the country. It is still standing, about half a mile north-east of Lake Mahopac, and owned by his son, the above-named Nathaniel.

Gen. Scott, with his Staff, made it his headquarters during a part of the Revolution.

EXTRACT FROM THE TOWN RECORD.

"At the First Town meeting held in the Town of Carmel at the house of John Crane Esqr. on the 7th of April 1795 The following persons were chosen for officers for said town, viz.:

    Robert Johnston Esqr., Moderator.
    John Crane, Esqr., Town Clerk.
    Timothy Carver, Supervisor.

    Daniel Cole,
    Devowe Bailey,   } Assessors.
    Thacher Hopkins,

    Elijah Douty, Junr., Collector & Constable.
    David Travis, Constable.

Devowe Bailey,
Daniel Cole, } Overseers of the Poor.

John Crane, Esqr.,
Timothy Crane,
Thacher Hopkins, } Commissioners of Highways.

Fence Viewers & Damage Prisers:

David Myrick,   John Berry,
Judah Kelly,   Samuel Jenkins,
Joseph Cole,   David Gregory,
Isaac Drew,   Billy Trowbridge.
John Crane, Esqr.,

Pound Masters:

James Townsend,   Stephen Fowler,
Joseph Crane,   Isaac Devine.
Wm. Webb,

Overseers of Highways:

Robert Hughson,   Israel Pinkney,
Seth Foster,   Peter Badeau,
Timothy Carver,   Gilbert Hunt,
John Bezea,   Gilbert Travis,
Joseph Hopkins,   Wm. Vermilyed,
Isaac Purdy,   Nathl. Boundig,
Philips Smith,   Job Austin,
Peter Maybie,   Gilbert Adams,
Thaddeous Raymond,   Stephen Crane,
David Gregory,   Joseph Cole,
John Cole,   Jeremiah Hughson,
David Frost,   Wm. Fowler,
Benjamin Crosby,   Abraham Everitt,
William Haden,   Daniel Thomas,
Jacob Ganung,   John Ganung,
Jonathan Whiten,   David Frost,
David Longwell,   James B——.

Voted that a hundred pounds be raised for the support of the poor the ensuing year. Voted that the next annual Town meeting be held at this place.

JOHN CRANE,
Town Clark."

There were 37 road districts in this town in 1795, laid out by the Commissioners of Highways.

"Whereas Joseph Gregory of the town of Carmel in the county of Dutchess and State of New York hath proposed to emancipate and Set free three female Negros the property of the said Joseph Gregory agreeable to a Law of this State in that case made and provided.

"We Robert Johnston & John Crane Esqrs. two of the peoples Justices of the peace for said county and Elisha Cole and Tracy Ballard Overseers of the poor of the town of Carmel do hereby Certify that we think that the said female Negroes That is one named Anglesse aged about 26 years one other 6 years named Rose and another named Dinah aged about 3 years are all sufficient to provide for themselves.

"Given under our hands this 3d day of January 1798.

"Robert Johnston, } Justices of
"John Crane,   } the Peace.

"Elijah Cole,   } Overseers of
"Tracy Bullard, } the Poor.

"John Crane, Town Clark."

Carmel Village.—A quiet, rural, and small village, beautifully situated on Shaw's Lake. The Courthouse, Jail, Clerk's Office, and Putnam County Bank, are located here. Through this village, in the olden time, ran one of the roads leading from the city of New York to Albany, and places in its vicinity. Five terms of the County Court, and General Sessions, and three terms of the Circuit, Oyer and Terminer Courts, are held here. The location is dry, elevated, and healthy. It contains 3 churches and 4 or 5 stores. It is named after the town in which it is located. It is 20 miles from Cold Spring, and 16 from Peekskill. A few rods north of the village, James Raymond, Esq., is erecting a family cemetery on a magnificent scale.

When completed, with avenues and walks laid out, gravelled, and ornamented with appropriate trees and flowering shrubs—with the *tree* of *Heaven*, the *Babylonian willow*, the dark funereal *yew*, and the mourning *cypress*—it will form a lovely and interesting addition to the suburbs of the village. The ancient taste for erecting rural cemeteries is reviving among us, and develops a chaste and holy feeling of our nature. He who cherishes a sacred regard for the dead, will prove an ornament to the living. All nations, Christian and Pagan, cherish a sacred regard for the last earthly home of those they love. The eastern nations selected the groves and recesses of wooded heights and secluded vales, beyond the city's serried wall, as places of interment.

By the Laws of the Twelve Tables, in the year 454, B.C., it was prohibited to bury within the city of Rome; and the Potters' Field was located without the walls of Jerusalem. The wealthy Israelites, we are told, built their tombs in the mountains near Jerusalem; and in a garden near the base of Calvary, Joseph of Arimathea had prepared that memorable sepulchre, in which was laid the body of the crucified Messiah. The Athenians permitted no burials within their city. In the gorges of the wooded hills on the opposite bank of the Nile, were the catacombs of Thebes, and beyond the lake of Acherusia were those of Memphis, from whence the Grecian mythologists derived their fabulous accounts of the Elysian fields.

Those illustrious men who fell in the battles of their country were buried in the Ceramicus—an extensive and beautifully ornamented public cemetery, where were the Academy and Gymnasium, with their superb gardens. Even the rudely built tumuli of the American Indians, reveal the tenacity with which they cling to the memory of their dead. The experimental garden and rural cemetery of Mount Auburn, at Cambridge, Massachusetts, may well vie with the celebrated Pere La Chaise, at Paris. The Congressional Burying Ground, at Washington, is another interesting spot, where man may "seek the living among the dead," and learn wisdom among those mute, silent, and melancholy memorials, that testify of his mortality here. It was Gray's Elegy, written in a country church-yard—

"Through which the ringing earth-worm creeps,"

which, more than any other of his writings, has given him a name and a fame in the literary world, that will survive and live, when brass and marble shall have crumbled to dust.

Since writing this article, we have been informed that this cemetery is to be called the "*Raymond Hill Cemetery.*"*

As this sheet is about passing from our hands to those of the compositor, we have only time to add, that a large and beautiful building is now being built in this village, and is to be known, as we are informed, as "*The Carmel Collegiate Institute.*"

The Telegraph line from New York to Albany runs through the village, having a station here for the accommodation of the villagers, and those residing in the vicinity.

Red Mills.—A small village situated on the Muscoot River, 8 miles south-west from Carmel. The

---

* As we go to press, we are informed by the Rev. H. G. Livingston, that it has been deeded to the Trustees of the Gilead Presbyterian Society.

mill and nearly all other buildings here are painted red; and this fact has given the name to the place. The first carding machine put up in this country permanently, was established here. It was brought to Peekskill by an Englishman named Ellinworth, about the year 1800. He put it up at Peekskill, where it remained about two years, and then removed it to this place, where it was looked upon with as much wonder and amazement, as the elephant, *old Tip*, was, when first exhibited in this county by ——, of the town of South-east.

It was supposed that English Custom-house officers were bribed to let it pass. Ellinworth brought a man by the name of Hague, who had worked it in England, to superintend its operations in this country. Hague returned to England, but Ellinworth did not dare to. It was about half as large as those made at the present time; and the Yankees were not slow in their improvements upon it.

Major Roger Morris, who married Mary Philipse, had a log mansion here. It stood about 40 rods northeast of the Post Office. Mary, or "*Madame Morris,*" as she was called by the tenants, was a remarkable woman, who possessed not only the esteem, but the love of the tenants of her estate. The "Long Lot," which Major Morris obtained with his wife, included the Red Mills. Major Morris and his lady lived a greater part of the year at New York and Harlem, and at a certain season would come up and spend some time at this place to receive rents and give directions to their tenants. The gentleness of manner and kindness of disposition manifested by Madam

Morris, soon secured the affections of the tenants; and not as yet having a house, in the estimation of the tenants, sufficiently large for her reception while she stayed among them, they got together and erected a large log-house, put up with more nicety, and finished in a style more suited to a lady of her rank and standing. She had ever been the friend of the tenants, and this act, on their part, was a tribute of their esteem and admiration for her noble and generous conduct to them. Isaac Lounsberry's house now encloses the log-house of Madam Morris. Major Morris is supposed to have built the first store and grist mill at this place, but at what period, we are not informed. Between where the log-house of Madam Morris stood and the Post Office is now kept, the mother of William Hill, now living, then a widow, and familiarly called "*Granny* Hill," lived in a log-house. This old lady had secured the friendship of *Madam Morris.* Some years before the Revolution, a kind of *anti-rent* rebellion broke out among the tenantry of the Morris estate. We are not advised of the true issue between the landlord and tenants, but believe it was somewhat similar to the anti-rent trouble of the present time. An *association paper* was soon drawn up and circulated among the tenants, making common cause with one another in the matter. Granny Hill asked to see the paper, and being deceived and misled as to the objects its signers proposed to secure to themselves, signed it.

The Major soon heard of it, and, calling on the old lady, required her to take her name off it. She, still believing that the paper contemplated nothing but what had been represented to her, refused to do so, alleging, as the story goes, that she "could seal it with her blood." She was told that she must then go out of her house; and out she went. The matter soon reached the ears of Madam Morris, who was informed of the deception played off on her aged tenant. She asked the Major what he had been doing with Granny Hill? He replied that she had signed "*that* paper, and had refused to take her name off; and that he had turned the old rebel out of doors." Madam Morris could not, for a moment, believe that the old woman would do anything wrong as her tenant, and somewhat resenting the hasty conduct of her husband, told him that "there was an 100 acres up the road with a log-house on it, and that Granny Hill should have a living on it for life;" gently reminding him that all the land was hers in her own right, and cautioning him not to molest the old lady again. Granny Hill accordingly received a life interest in the 100 acres; and after the estate was confiscated, in consequence of Major Morris taking the British side of the question, under whose government he held a commission, an individual who had purchased it of the State, demanded an absolute and full title; but it was found that the old lady had an interest in the land that could not be terminated, without her consent, until her death. When Gen. Montcalm, the Commander of the French Army, was shot in the battle on the Plains of Abraham, at Quebec, he was riding a large and beautiful white horse. The horse was captured by the British, and made a present to Major Roger Morris, for the gallantry he evinced in that battle, who brought him, after the war was finished, to the Red

Mills. He was kept on the farm adjoining Nathaniel Crane's, Esq., by George Hughson, the ancestor of the Hughson family, in Dutchess County, at that time an agent of Major Morris. The stock of this horse was found here within the last fifty years.

Indian Hill.—This is a large eminence at the south end of Lake Mahopac, cultivated to the top on the south side, and lately owned by Abel Smith. The Mahopac tribe of Indians occupied this region of the country; and hence the name.

Watermelon Hill.—This hill is about one and a half miles south-east of Lake Mahopac. The north side is still covered with timber; the south goes off with a gentle slope, and is cultivated. It is partly owned by Richard Dean, Esq. About 130 years since, a great hunter from New Rochelle, Westchester Co., called Captain Simpkins, came up here and found watermelons in great plenty on this hill; whereupon he named it as above. He was on friendly terms with the Indians who lived there, and with whom he bartered. In the Revolution, the cow-boys and horse-thieves built pens on this hill, in which they put stolen horses, until they could safely convey them to New York for the use of the British Army. About 20 years ago the remains of these pens were still to be seen.

Battle Hill.—This hill is in the southerly part of the town, on the lands of the —— Ganongs, Esq., two and a half miles south of Carmel village. It was formerly a great resort of rattle-snakes. Dropping the latter name, the people in its vicinity named it as above. A young man was shot on this hill in Revolution, and although found with a gang of horse-thieves, he was innocent of any participation in their nefarious deeds. He lived in the town of Pawlings, Dutchess County, had lately married, and was going to see his wife, who, at that time, was with her friends in the town of Bedford, Westchester County. The gang, who had their head-quarters at Pawlings, persuaded him to defer his journey for a day or two, by offering the use of one of their horses to ride, as they were going in the same direction to one of the American posts on the Lines. He accepted their offer, not doubting their representations about the horses being for the American army. They had proceeded as far as this Hill, where, encamping for the night, they were overtaken and attacked by the owners of the horses and their neighbors. The gang escaped, but the young man, as he rose up from the ground beneath a tree where he was sleeping, was shot through the back. He died in 48 hours afterwards, but lived long enough to see his wife, who was sent for, and explain to her and those around him, how he happened to be found in such company. He was buried a few rods north of the hill.

Drew's Hill is a large eminence directly east of Rattle Hill, on the land of John Craft, Esq., and named after the Drew family.

Pond Hill is about two miles south of Carmel village, at the north side of the Gilead Pond, at its foot.

Watts' Hill.—A small eminence just east of Pond Hill, on the farm of Judge Watts, after whom it is named.

Hazen Hill is about one and a half miles southwest of Carmel village; and named after the Hazen family, who were among the earliest settlers from Cape Cod.

Berry Mountain.—This is a large eminence, named after Jabez Berry, to whom it formerly belonged, and is now owned by Messrs. Wixon and Ballard. On its summit is a tree, from the top of which, when the woods are destitute of foliage, seven fish ponds can be seen, five of which are visible at all times of the year.

Hitchcock Hill, now called Prospect Hill, is about two and a half miles north of the Red Mills. It has been in the possession of the above-named family for a hundred years.

Round Mountain is about one and a half miles east of Hitchcock Hill, owned by Messrs. Hill, Pinckney, and Barrett. The shape of this eminence is circular; hence the title.

Turkey Mountain is about one mile east of Round Mountain. Formerly it was covered with white-oak timber, and frequented by wild turkies.

Corner's Mountain.—A small eminence about one and a half miles west of Carmel village, formerly owned by a person of that name.

Austin Hill is about two and a half miles northwest of the Red Mills. Job Austin's father, who came from Germany about 100 years ago, settled on it. It has always been in the possession of the family.

Big Hill.—This is the largest and highest ridge of land on the east of the Peekskill Hollow range of the Highlands, and situated about two and a half miles south-west of the Red Mills.

Lake Mahopac.—This well-nigh unrivalled, beautiful, and romantic lake, is located in the westerly part of the town, thirteen miles from Peekskill, Westchester Co.; five from Croton Falls in the same county; and four miles south-west of Carmel village. It is a delightful watering place, crowded to overflowing in the summer season with visitors from New York City and all parts of the country. During the fashionable season " strangers arrive here every day, and those who have been before, never go anywhere else for recreation and enjoyment, for they can be had here perfectly unadulterated. This Lake is nine miles in circumference, and is situated about eighteen hundred feet above the level of the sea. It is one of the principal sources of supply to the Croton; and its pure and placid waters, its wide and picturesque scenery, the romantic resorts, its wild and wooded islands, the frequent and agreeable *pic-nic* parties to the *Dell*, with its clear and crystal spring, its rugged and precipitous cliffs, and last, but not least, the rowing and sailing among the islands and along the wild and rock-bound shore are delightful, and superior to anything of the kind in any place it has ever been our good fortune to visit. Kirk ridge has an attraction for the curious, in the position it occupies between two lakes, one of which is about 150 feet below the other, both of which can be seen on this shore.

"This Lake affords fine sport to the angler. Pickerel, pike, and perch are caught in abundance, and its shores abound in all the wild game of the season, particularly woodcock, affording excellent sport to those fond of gunning. There is one peculiar feature about this lake, which we have never found in any other watering place on this continent, and that is the complete absence of all desire for artificial amusements, such as bowling, billiards, cards, games, and all those things, which sojourners at such places usually resort to for the purpose of killing time. Nothing of this kind is required here; the natural resources of the place are sufficient to keep the most attractive constantly engaged, and it would take weeks and months to exhaust all the facilities the place possesses for amusement.

"In all watering places, much depends upon the society and the class of people there congregated. So far as this important matter is concerned, Lake Mahopac is particularly favored. There is no exclusiveness, no coteries, no codfish aristocracy, but all appear disposed to make time pass as agreeably to those around them as to themselves.

"Our watering places, generally, are the rendezvous of fashion, and there are usually more restrictions upon dress, and more restraint upon personal movements than even in cities. The benefits derived from the cool air, and the relief from care, a resort to the country usually gives, are not enjoyed, and everything is sacrificed to fashionable dress and fashionable hours, more so, if possible, than in the most fashionable city. Sensible people do not go to such places for such purposes; it would be more sensible to stay at home; but they go where they can get all the comforts of the country, where they can enjoy themselves in their own way, or in other words, do just as they please; for such purposes, they come to Lake Mahopac. It is a fine place for young children, on account of the facilities for bathing, the pleasant drives and walks.

"It is a very desirable resort for all, as it is easy of access, being only five miles from the present termination of the Harlem Railroad at Croton Falls, and only three hours' ride from the city. The ride to the Lake from the railroad is beautiful, it being through such a wild and picturesque country, and is really refreshing after fifty-four miles of railroad travel.

"There are more beautiful places for summer rendezvous in the vicinity of the city of New York, than in any other in the Union, and as fast as our railroads become extended, new resorts are opened. It is only within a few years that this place was known to any extent by the citizens of N. Y. City, and we predict that within a few years, it will be more generally resorted to than many places which have been longer and hitherto more favorably known."

Kirk Pond.—This handsome body of water is one mile long, half a mile wide, bounded by the lands of Messrs. Hill and Lounsberry. It takes its name from an old man by the name of Kirk, who lived near it; and abounds in excellent fish.

Wixon Pond.—A large pond about half a mile north of Lake Mahopac, nearly circular, and containing excellent fish. This and Seacord's pond are the only ones in the county that originally contained white perch. On the third day of December, 1838, Nathaniel Crane, Esq., with a scoop-net as large as a half bushel, scooped out of this pond eleven bushels of white perch.

Long Pond is bounded by the lands of John Wixon, Allen Coles, Ebenezer Barret, and Alza Hill, Esqs.

Cranberry Pond is south of Nathaniel Crane's, Esq.; bounded by the lands of Lewis Griffin, Reuben Baldwin, and Coleman Rockwell, Esqs. It covers about thirty acres, and contains perch, pike, with the more common kinds of fish. Cranberries are found in abundance on its borders.

Shaw's Lake.—This beautiful sheet of water is sometimes called Shaw's Pond, on the east and north banks of which is located the quiet little village of Carmel. Its location is a basin, as it were, scooped out of the surrounding hills. It is about one mile in length, three quarters in breadth, and 130 feet in depth, containing all kinds of fish in great abundance. At its north end, on elevated ground, is the charming residence of James Raymond, Esq., the main proprietor of the largest collection of wild animals ever exhibited in this country. At the south end, on a still higher eminence, stands the former mansion of Samuel Gouverneur, deceased.

At the north end, a man by the name of Shaw resided before the Revolution, after whom this Pond or Lake was named. We have been informed that the villagers of Carmel are about adopting a new name, by which this noble body of water shall, in future, be called; but, as yet, they have not published it *pro forma*.

Gilead Pond, formerly so called from its contiguity to the old Gilead Church, is now known as the Crosby Pond; and is situated about one mile south of Carmel Village. It is nearly a mile in length, half a mile in breadth; and, like Shaw's, abounds in all kinds of fish. It takes its later name from a son of the celebrated Enoch Crosby, who owned a mill, which stood on its out-let.

Barrett's Pond is in the north part of the town, near the line between it and Kent, covering about ten acres of ground. It bounds on the lands of different members of the Barret family, after which it is named.

Seacord's Pond, is half a mile wide, circular, and named after the Seacord family, who lived close to it.

Capt. John Crane.—This gentleman, the father of Nathaniel Crane, Esq., now living in this town, was born the 20th of Nov., 1742, old style. He built the house, now occupied by his son Nathaniel, in 1772. His ancestors were among the earliest settlers in this county, of English origin, and of great influence and intelligence, wherever located. * During the Revolution, the Crane family figured largely in those trying times "that tried mens' souls," both in the civil and military departments of the government.

Nearly all of the name, in this country, have descended from John Crane, who came from Suffolk county, in England, about 1675, and settled in Massachusetts. He fought in the Indian war of 1720, at Deerfield, and was in the fort when taken by the Indians. By making a passage under the logs, he succeeded in escaping with his family; and afterwards settled at Wilton, in Connecticut. He had two sons, Jonathan and Jasper. Jasper settled at Elizabethtown, in New Jersey, and was the grandfather of Col. John Crane, of the artillery, in the Revolution. Jonathan had one son, Joseph, who was born 17th of May, 1696. Jonathan settled in Massachusetts: and his son Joseph, grandfather to John Crane, came from Greenfield, in Connecticut, about 1755, and settled on the north side of Joe's Hill, about one and a half miles east of Sodom Crane. He built the mill there, called in the old records, "Crane's Mill."

Another branch of this family, Orrin and Anson Crane, Esqs., now living in the town of South-east, are grandsons of Joseph Crane.

The branches of this family, in this town and Southeast, have kept regular chronological histories of the family; and intending to insert them, we pass over them to resume our remarks concerning that member of the family at the head of our article. The whole family seems to have been distinguished for integrity, intelligence, and attachment to the cause of their country.

In searching the continental, provincial, and military records of the Revolution, we have not found one of the name, adhering to the cause of England; they were all whigs at that day, and thoroughly "dyed in the wool."

Joseph Crane, uncle to Jonathan, the father of Anson and Orrin, was a Colonel of the Militia in the Revolution, and fought in the battle at Ridgefield, when partly destroyed by the British, after burning Danbury. John, the subject of the present sketch, was a Captain, and in 1803, '4 and '5, was an assistant Judge of the Court of Common Pleas, of Dutchess County. All of both branches of the family in New York, as well as in New Jersey, who were old enough to bear arms, held commissions, either in the Continental army, the militia, or in minute-men companies. John seems to have become early the subject of hate and fear to the tories and friends of the King. Attempts were made to capture him in his own house when alone, and shoot him when out of it; but their efforts were foiled by a Power that watched over the American cause and its advocates. In the fall of 1780, he retired to bed with his wife, having carefully secured the doors and windows. About an hour afterwards he heard a rap on the side of the house. He arose and looked out of the window, which was half boarded up, and told his wife that he saw two men, armed, outside. It was moonlight, and the refraction of the moon's rays gave the appearance of two men, when in reality there was but one. A reward of $200 had been offered for his apprehension, which he supposed had induced a band of tories, who lived in the vicinity, to pay him a night visit. He supposed that there were others hard-by, secreted behind trees and fences. He slipped out of the back door, cautioning his wife to fasten it, intending to secrete himself in an adjoining wood. After his departure, his wife, on looking out of the window, saw but one man. The man spoke to her, and begged the privilege to come in and get something to eat, and rest himself for the night on the floor. She asked him if there were others with him, and he assured her, on the honor of a soldier, that he was alone. She then asked him if he was armed? He said he was, and that "Washington's soldiers always went armed." She took off the fastenings and raised the window a little, and told him to hand in his gun breech foremost, which he did; and having fastened it again, she cocked the gun, opened the door, and bid him come in, standing a few feet back, ready to shoot him and close the door should another make his appearance. She then bade him fasten it, and having placed the gun in a corner, got the soldier some supper. While he was eating, Crane crept up to a back window, and seeing but one man in the room, and he quietly eating his supper, called to his wife to let him in again. He came in and begged his wife to say nothing about his flight from one man, since it had turned out so differently from what he expected; but she declared that it was too good to keep, and many a day afterwards she rehearsed it before him to the no small amusement of his friends.

One night previous to this, while sitting by his fire reading, and his wife in the corner darning stockings, a cow-boy and tory of the name of Samuel Akerly, of South-east, came to his window with gun in hand, intending to shoot him. Akerly contemplated the scene within, remembered the former friendship and kindness he had received from Crane, (long before the *great issue* was joined, that arrayed neighbor against neighbor,) and withdrew, afterwards alleging that "Crane was so great a friend to his country, and so sincere in his actions, that he could not do it." At another time Capt. Crane went to his field, a few rods west of Nathaniel Crane's house, now a meadow, to chain his horses for the night. Akerly was lying in wait to shoot him. Again he suspended his purpose, alleging that his "heart failed him."

About this time, Robert Hughson, a whig and neighbor of Capt. Crane's, went out one night on the ridge just east of Crane's house, and was met by three horsemen, well armed, who enquired, whether one Capt. John Crane, did not live in the house, to which they pointed? Hughson told them that he did. They then told him that he must go with them and assist them to get $100 dollars in hard money, which they said Crane had concealed in a bin of grain in the upper part of an old log-house, just back of his dwelling. Hughson told them that if he must be shot, he would rather it should be on his own land, than on the door-sill of his neighbor, among a band of robbers; that

Capt. Crane had four men with him well armed, and that before they could get the money, some of them would have to "*bite* the dust." Hughson so magnified the force and fighting disposition of Capt. Crane, and the danger of their all being killed before they could get into the house, that they departed without making an attempt.

The family Coat of Arms is now in the possession of Nathaniel, his son, and a *Record*, reaching from the great ancestor of this family, John Crane, of England, down to the children of the subject of this biography, and written by himself when he only lacked six days to being 83 years of age. We extract it entire, considering it wrong to comply with his request, so modestly expressed at its end. A majority of the old men of the present time, who have reached his age, can hardly do more than write their name, while many of them can, at the most, only make their marks. We doubt if there is one in a thousand, who can write four large quarto pages, with the penmanship anything like as good, as that which now lays before us, in the handwriting of Capt. John Crane:

A RECORD OF THE CRANE FAMILY.

"Carmel 29th Nov'r 1825. the following is a Reckord of the Crane family as handed to me by my ancestors—my Grand-father Joseph Crane was the son of Jonathan Crane of Windham in Conecticut and grandson to John Crane from England and was born May 17th 1696.

" Died August 28th 1781.

" Zebulon his eldest child was born Jan. 25 1721 Died Jan 24th 1789

" Joseph his 2d child was born Sep 13th 1722 Died Oct 14th 1800.

"Mary his 3d child was born May 30th 1726 Died March 17th 1805.

"Thadeous his 4th child was born March 28th, 1728 Died Septr. 1803.

"Abijah 5th Child was born April 3d 1730 Died 3d 1806. Anna 6th child Born April 12th 1732 Died March 25th 1805. Stephen 7th Child Born May 13th 1734 Died May 10th 1814. Adah 8th and last Child born Oct 25th 1736 her deth I dont remember, but she lived about 70 years. here ends the record of my grand-father's family.

"here begineth a record of my Father's family—I John Crane was born Nov 24th 1742 old stile—William was Born 1744. Zebulon was Born August 7th 1746 Died December 31st 1814. Elijah Born April 1st 1748. Sarah Born July 12th 1750. Mary Oct. 8 1752—Stile altered about this time—Belden Born Nov 31 1754. Samuel Born April 11th 1757. Abigah Born May 26th 1759. Stephen born April 11th 1761. Anna born Augt. 3d 1763. Seth born march 1766. my mother was Sarah Belden— Before mared Daughter of Wm. Belden of Wilton in Conecticut and was a resident of the town of Dearfield in Snoserjoseraets at the time it was Destroyed by the French and Indians in the winter of 1720 & 3—

"in the later part of the year 1769 my Father moved from Bedford in Westchester County to Judeah now in the town of Washington in Litchfield county—Soon after they Got there a mortal sickness Came into the family, in which my mother and five of her children died within 2 months, namely—Mary, Belden, Stephen, Seth, anne—My mother had never lost a child before—

"I John Crane was born Nov 24th old stile 1742. My wife Tamer, Daughter of John and Hannah Carpenter, was born Dec'r 1st 1747, Died May 1st 1823. we were married on the 1st day of March 1764 By the Rev'd Eliphelet Ball the first settlar and Minister of Ballstown in the State of New York—our Eldest Child Joseph was Born June 3d 1766.- 2d Child Adah was born June 6th 1768. 3d Child Stephen Nov 1st 1770, Died Sept 9th 1826. 4th Child John born June 6th 1773, Died June 1st 1825. 5th Child Zillah was born Oct 3d 1775. 6th

Child Nathaniel born Feb. 28th 1778. 7th Child Sarah born June 27th 1780. 8th Child Arrabellah born 25 Decem 1784. 9th and last Child Clorinda was born Oct 2d 1787.

"I hope it will be a satisfaction to some of my Descendants to be informed of the conduct of their ancestors throu life—my Grand-father was Living at the commencement of the Revolutionary war that separated the then 13 Collynies from the Government of Grate Brittan—at the Commencement of that war the People were devided into two Casses whig & tory—the whig party ware those opposed to the black arts of the British Parliament—the tory party took sides with the King—my grandfather was then about 80 years of age very strong and active for a man of that age, and a warm whig and what is very remarkable, his 8 children were all living and heds of familyes, had many grandchildren and great grandchildren and not an individual that had arrived to the years of understanding but what took an active part of the American Cause—I was the oldest grandchild—I had an Ensign's Commission under the King George the 3d in the year 1775—

"I took a Capt's Commission under the Provential Congress of the Provence of New York—the 4th of July following our independence was declared—George Clinton Became our Governor—then I received a Commission from him and held it through the war—Such was the general conduct of the family which was the cause of many of them receiving both civil and military commissions, not on account of our Extraordinary abilities, but as an act of our engagedness in that blessed cause—I hope whoever reads the foregoing will Erase the incorrectness, as I want but *six days* of being 83 years of age and allmost blind.

" JOHN CRANE."

John Crane was a remarkable man ; and the record he penned, at the age of 83, is very far from diminishing the force of our assertion. He was a kind neighbor and indulgent parent, a firm friend, and unflinching patriot. He died in 1825.

Lieut. Jabez Berry.—This gentleman's ancestors

were from Ireland, and emigrated early to Cape Cod in Massachusetts, where the subject of this memoir was born. We have been able to gather but little concerning his early life, previous to his arrival in this country. His ancestors, himself, and his descendants, were, and are still, distinguished for their gigantic proportions, muscular frames, and great strength. Jabez Berry came from Cape Cod after he was married, and some years before the Revolution, and settled on the farm now occupied by Elijah Crane, about one mile north of Lake Mahopac. He was five feet eleven inches high in his stocking feet; a large, powerful, robust man, with a frame knit together more like iron than bone, and capable of the greatest endurance. For his size, he was unmatched in strength by any man, at that time, in the country. *Boxing* was one of the amusements of the young men at that day, very fashionable, and, as a science, is still cultivated. He soon attained great proficiency in it, and before leaving Cape Cod stood number one, "solitary and alone," and without a rival. Some years after settling in this town, a celebrated *boxer* came to Cape Cod and inquired for one Jabez Berry? On ascertaining that he had removed to this town, he informed one of Berry's intimate friends there that he came to have a *match* with him, and offered to bet that he could flog him. Berry's friend, well-knowing his ability, accepted the wager; and another person having been chosen as the second of the boasting bully, the three immediately set out for Berry's residence. On reaching it they found him and his wife at breakfast. The boxer without much ceremony entered the house, and thus accosted Berry: "Are you the man they call Jabez Berry?" "Yes sir-ee, and always have been," was the reply. "Well, sir," continued the bully, "I have come all the way from Cape Cod to flog you." "Ah, indeed! If you've come all that distance to pluck a single *berry* from the bush, you are entitled to a few *striking tokens* of my regard as a reward for the *pains* you may suffer before you get back," was the reply. Out they went into the door-yard, where he flogged his Cape Cod antagonist to his heart's content; received half the bet, which he applied to curing his antagonist, who was unable to resume his journey back for the space of a week. There was one remarkable trait about him that distinguished him from others who possessed great powers and skill in pugilism; he never made use of it to domineer over the weak and those unable to cope with him, nor insult any man from a consciousness that his skill and strength was his protection from punishment. He never was the assailing party; nor entered a boxing combat in an angry state of feeling. He enjoyed it with about the same good feeling that he would relate an amusing anecdote or crack a harmless joke. He belonged to the church; and if sickness or bad weather did not prevent him, never failed in his attendance for any other cause. He was commissioned a lieutenant in the militia, and rendered great service in guarding this part of the country from the midnight depredations of the cow-boys, skinners, and tories. He had four sons; John, who was commissioned an ensign in Captain John Crane's militia company, Asahel, Jabez, junior, and Samuel, and two daughters, His sons are all dead, but some of his grandchildren are found in this town. Samuel A.

Berry, Esq., of Carmel, whose father was Samuel, the youngest son of Jabez, is a grandson of this early settler. Samuel had four sons; Charles, John, Frederick, Samuel A.; and seven daughters, Delilah, Hester, Elizabeth, two by the name of Clarissa, one having died young, Julia and Mary.

Jabez Berry possessed a well-balanced mind, which kept him from being disconcerted in any emergency. Possessing an amiable and cheerful disposition, he secured the esteem and approbation of all who knew him, while his integrity and uprightness of purpose secured him from the tongue of the slanderer. He advocated the cause of his country with a stout heart and a strong arm, and enjoyed the proud satisfaction of seeing all of his sons follow his paternal and patriotic example.

We have not been informed of the date of his decease, but he lived many years to enjoy the fruit of the tree of liberty which he had contributed so vigilantly to guard.

We had intended giving a brief sketch of the Churches in the county; but on inquiry found that they were principally of recent organization, not more than three being organized at the commencement of the Revolution.

The first Church erected in this county, so far as we are informed, was built about 1735, in Southeast, in which, about 1740, the Rev. Elisha Kent, the grandfather of the late Chancellor Kent, preached as the regular pastor.

St. Philips' Chapel, as it was then called, is the Episcopal Church near the Hon. John Garrison's, and was built in 1770, by Col. Beverly Robinson.

During the Revolution, it was used as a kind of Jail to confine prisoners. One minister preached here every other Sabbath, and also in a Church just south of Continental Village, across the Westchester County line, where Col. Beverly Robinson gave the two Churches a farm of about 300 acres as a parsonage. A Church was built previous to the Revolution in Patterson, but we have not obtained the necessary facts to give a sketch of its history.

We had abandoned the idea, therefore, of saying anything concerning the Churches, inasmuch as we were unable to notice all, but could not forego the pleasure of inserting the following notice of the Gilead Church, which has been politely furnished us by the Rev. Henry G. Livingston, its present *Pastor*.

THE GILEAD PRESBYTERIAN CHURCH.

About the commencement of the Revolution, a Congregational Church was organized in the vicinity of Carmel Village, and a log building erected on the hill a few rods north of the present residence of Ira White, Esq., and within the limits of the town of Southeast. The Society was familiarly known as "Gregory's Parish," after the name of their first minister. No authentic records of the Church are found until 1792, when a new organization was made, and a more commodious edifice built upon the ground now known as the Gilead Burying-ground, a little over a mile south of Carmel Village.

The Constitution and Articles of Faith, then adopted, are as follows:

"Frederickstown, August 9th, 1792.
"We, the subscribers, members of different churches, and of the

former church in this place, now dissolved, living in the vicinity commonly known as Gregory's Parish, considering it the duty of Christians to join together in covenant, and form churches for the glory of God and their mutual edification, wherever God in his providence may cast their lot, and place them under circumstances convenient for that purpose; and finding ourselves under such circumstances, and no church in this parish which we may join, and with which we can walk in the ordinances of the Gospel according to our persuasion; and having, as we humbly trust, looked to the Father of lights for wisdom and direction, and having also consulted with ministers and private Christians concerning our duty under present circumstances, have, after mature deliberation, judged that we ought, with the consent of the churches to which we belong, to unite together in covenant as a visible church, and Messrs. Ichabod Lewis, John Minor, Amzi Lewis, and Silas Comfort, Ministers of the Gospel, having by our request convened in order to assist us to unite and enter into covenant with each other with solemnity and propriety, we have therefore adopted and publicly received the following articles and covenant as the foundation of our union:—

"Articles of Faith.

"1. There is one only living, true, and eternal God, the Creator, Preserver, and Governor of the world; infinite in all perfection and glory, and worthy to be loved, worshipped, and obeyed by all rational creatures.

"2. The Scriptures of the Old and New Testaments are the Word of God, and a sufficient and infallible rule of practice.

"3. Mankind are fallen from God, and are naturally destitute of all holiness and inclined wholly to sin; and therefore are under the curse of God's law, and deserve his eternal wrath.

"4. God purposed in himself before the foundation of the world, to save some of the human race by a dispensation of grace through a Mediator.

"5. This grace has been revealed to fallen man by Jesus Christ, who, being really God, became man, and in the flesh performed the work of mediation, and by his obedience and death, opened up a way in which sinners may be freely justified by faith, and saved according to the divine purpose.

"6. The Mediator, Jesus Christ, is the appointed Governor of the world and the final judge of the quick and the dead.

"7. Those and those only who are chosen of God in Christ, and renewed by the effectual operation of the Holy Ghost on their hearts, do actually repent and believe unto eternal life.

"8. God will continue his gracious operations on the hearts of his people until they are completely sanctified and fitted for his Heavenly Kingdom and glory.

"9. God will overrule all things for his glory and the advancement of his Kingdom until the consummation, when those who are united to Christ by faith will be raised and glorified, and the impenitent and unbelieving eternally punished."

"*The Covenant.*

"We do this day solemnly take God for our God, Jesus Christ for our Saviour, the Holy Ghost for our Sanctifier, and the Scriptures for our rule and directory; and sincerely, as far as we know ourselves, covenant and engage by divine grace to devote ourselves to the service and glory of God, walking in all his ordinances, observing his commandments, living solely, righteously, and godly in this present world, trusting in the merits of Jesus Christ alone for acceptance with God, seeking his Glory and Kingdom, watching over our Christian brethren and sisters in love, studying to promote their spiritual edification, and therefore good, endeavouring to keep the unity of the spirit in the bond of peace, and waiting for the coming of our Lord and Saviour Jesus Christ.

Signed,

John Ambler,	Mary Hopkins,
Matthew Beale,	Desire Stone,
Philetus Phillips,	Mary Haynes,
Zebulon Phillips,	Lucy Cullen,
John Merrick,	Bethia Trusdell,
John McClean,	Esther Phillips,
Jabez Trusdell,	Elizabeth Merrick."
Rebecca Hopkins,	

Who were the officers appointed in 1792, does not appear from the minutes of the Church record.

It is believed there was then no settled minister, though the pulpit was occasionally supplied by the Rev. Mr. Lewis. At a meeting held, December 9th, of the same year, it was resolved that the name of the Society be changed from that of "Gregory's Parish," to that of "*Gilead.*"

Since then, it has always been known as the "Gilead Church." The name was derived from Scripture without any special reason, I apprehend, for so appropriating it. Fancy, doubtless, dictated the selection.

June 27th, 1795, Mr. John Amber was elected deacon of the Church, and the first who filled that office.

Soon after this period, the Church gradually lost strength; there being no settled pastor, and, in all probability, no stated administration of the ordinances. It was, therefore, thought advisable, in the year 1803, to reorganize. The same constitution and articles of faith and government, as before, were adopted, and were signed by the following individuals:

Elisha Smith,	Thirza Crosby,
Dorius Crosby,	William Jacks,
Harvey Newell,	David Travis,
Gilbert Travis,	Rachel Newell,
Elizabeth Travis,	Jane Woodhull,
Desire Stone,	Deborah Travis,
Denny Jacks,	Hannah Rimdee.

Harvey Newell, at this time, was elected Clerk, and William Jacks, deacon. The Rev. Mr. Stephen Dodd had been previously called to the charge of the church, and preached here half of the time, and the remainder at the Red Mills.

He remained here until the 15th of July, 1810. During his ministry, a large number were added to the church, and its condition was prosperous. He was not, however, a settled pastor.

February 18th, 1804, Enoch Crosby, the supposed hero of Cooper's novel, entitled the *Spy*, and so well known for the aid he rendered his country in its time of trial, was elected deacon; and, in 1806, David Travis was set apart to the same office.

After Mr. Dodd had resigned the care of the church, the pulpit was supplied by the Rev. Herman Dagget, who was succeeded in 1812, by the Rev. Allen Blair. Afterwards, the Rev. Messrs. James N. Austin, Abner Brundige, Isaac Allerton, and B. Y. Morse, officiated from 1815, until the year 1835. None of these, so far as my knowledge extends, were settled as pastors.

The church, as it appears from the minutes, from the year 1824 to 1831, was in a state of fearful declension. The ecclesiastical body to which it belonged, was gradually becoming extinct. Disorder and a relaxation of dicipline, naturally resulted from this state of things. The preaching of the *Word* was feebly sustained, and but few were added to the congregation of the Lord. From 1831 to 1835, there was stated preaching every Sabbath; but the church had little more than a name to live, although there were set times when Zion was far advanced, and some few made public profession of their faith. In March, 1834, the church made the following declaration:

"We, the members of the Second Presbyterian Congregational Church in the town of Carmel, and formerly having been a branch of the Westchester Presbytery, which is now extinct, do declare ourselves to be, as in fact we are, an independent Congregational Church. Believing, however, that great benefits may result to the Church of Christ from intimate union and fellowship with each other by their mutual aid and counsel, hold ourselves willing to unite with some ecclesiastical body whenever, in the providence of God, an opportunity shall present and the way made clear."

The church, soon after this, assumed the Presbyterian form of government, and connected itself, June 3d, 1835, with the Presbytery of Bedford. Joseph Crane, Gilbert H. Travis, and Morgan L. Raymond, were elected Elders. In October, of the same year, a call was extended to the Rev. Gilbert Livingston Smith, to become pastor of the church. The call was accepted, but before he had entered upon the active duties of his office, he was suddenly translated to the Church of the redeemed above.

In the year 1837, the society erected their present house of worship, in the village of Carmel; a structure justly admired for the neatness of its finish and the beauty of its location. The Rev. G. T. Todd was soon after installed pastor, and the first who was ever settled as such; and remained until May, 1844. He was succeeded in August, 1845, by the Rev. Henry G. Livingston, who was then ordained and installed, and continues to fulfil the duties of the pastoral office at the present time.

The church has gradually increased since 1835, and has now upwards of 100 members. Though its growth is slow, it is sure, and its friends, to whom its past history is familiar, have every reason "to thank God and take courage."

The present officers of the church are, Rev. Henry G. Livingston, Pastor; Gilbert H. Travis, Morgan L. Raymond, Daniel Travis, and Anson Fowler, ruling Elders. Its creed is the same as that of other orthodox Presbyterian churches; and its ecclesiastical relations are with the Presbytery of Bedford, and the Synod of New York.

TOWN OF SOUTHEAST.

This town was organized in 1788, and, as before remarked, takes its name from its geographical position, being located in the south-east part of the county. Its surface is rolling, and indented with vales and low lands, which yield excellent grass; and, as a town, is better adapted for grazing than grain. The Harlem Railroad is now being made, nearly through its centre, from south to north, and will greatly facilitate the transportation of its farmers' produce to market.

The main and east branches of the Croton traverse it from north to south, and with a few other streams, furnish sufficient water-power for milling purposes. Nowhere in the County, except in Patterson, have we found better roads, at least in that part through which we rode.

Agriculture, as a science, seems here to have kept pace with every other; and the farm-houses, with their out-buildings, give a pleasing evidence of neatness, thrift, and good husbandry. Its population in 1840 was 1910; in 1845, 2044.

EARLY SETTLEMENT.

This town was one of the earliest settled in the county. Adjoining Westchester on the south, and Connecticut on the east, emigration, as it flowed northward from the City of New York, and westward from Massachusetts and Connecticut, poured its tide into this town, Carmel, and Patterson.

There were a few families who came from Westchester and settled here, but the greater part were from the then Colonies of Massachusetts and Connecticut.

The rich fat lands of the Croton early attracted the attention of the citizens of the above-named Colonies, one generation of whom had worn themselves out in their attempt to subdue the rough and stony surface whereon they first settled.

The principal settlers of this town were the Cranes, Crosbys, Halls, Moodys, Paddocks, Hanes, Howes, Carpenters, and Dickinsons. Deacon Moody, as he was familiarly called, was about the first settler at Sodom Corners. He bought all the land in its immediate vicinity. A short distance north of him, James Dickinson made an early settlement. His son, James Dickinson, jun., was one of the Commissioners for laying out roads in 1745, with Thomas Davenport, Esq., of Nelsonville, in Philipstown.

David Paddock, grandfather to Mrs. Richards, now living in this town, came from Cape Cod, in Massachusetts, with a family of eight children, about 1740, and located near the Presbyterian Meeting-house. His children were, Nathan, Foster, David, Isaac, Mary, Susanna, Mercy, and Sarah. Isaac was killed in the fight at Ward's house below White Plains, in Westchester County; and when shot, he fell against Capt. Joshua Barnum, the grandfather of Col. Reuben D. Barnum, Clerk of this County.

While falling, his tobacco-box fell out of his vest

pocket, which Captain Barnum picked up and eventually returned to the father of young Paddock.

Nathan, after the Revolution, caught the small-pox, and died at Catskill, in Greene County; Foster died in Dorset, in Vermont; and David died in this town after the Revolution.

Caleb Carpenter, great-grandfather of the Hon. Azor B. Crane, of Carmel, on his mother's side, with twelve others, came to this town about 1730, and located about three miles north of Sodom Corners, where they built the old Presbyterian log Church, in which the Rev. Elisha Kent, grandfather to the late Chancellor Kent, first preached in this town.

Joseph Crane, grandfather of Ansin and Orrin Crane, Esqs., came about the same time, and settled on the north side of Joe's Hill, one and a half miles east of Sodom Corners, where he built the mill known in early times as "Crane's Mill."

Our article on early settlement is necessarily general and imperfect, as the sources from which we obtained it are limited; and, resting in the memory of a few aged people, they, in some instances, are not competent to give it with any great degree of accuracy.

Extracts from the Town Records.

"At a town meeting held at the South precinct in Dutchess County 6th day of April 1773
1 John rider Was chosen Moderator
2 Isaac Elwell Clerk
3 chosen Joseph Crane Jr. Supervisor
4 was Chosen John field Sessor
5 was Chosen Samuel Bangs Sessor
6 was Chosen peter hall Collector
7 was Chosen Thomas trowbridge Constable
8 mark Gage Constable

Joseph Hull poormaster
Zebedee brigs poormaster
Daniel haviland poormaster
Thomas baldwin } Commissioners for
Oliver he   } the highways

Seth Nicknerson Commissioners
Benjamin Sears pound Keeper
Daniel haviland pound Keeper
Nathan Goreen Jr fence viewer
William Stone fence viewer
Uriah Townsend highway master N 1
peter hall path master N 2
Nathan Green is path master N 3
William Penny Jr path master N 4
hervey hopkins path master N 5
Zebede brigs path master N 6
Nathaniel foster path master N 7

"*Births in South East.*

Mercey Clinton Was Born August the 31 1766
Phebe Clinton was born May the 24 day 1768
Estr Clinton was born May the 24 day 1770
Jesse Clinton Was born July the 21 day 1772
Joshua Hinkley was born March 11th 1775
Elkane hinkley was born July 19 day 1759

" Benjamin Tounsels Ears mark is a Crop on the Right Ear & A Nick under it & half penny on the Under Side of the Left Ear

" Isaac Elwell Ear mark is a Crop of the Left with a hole in the same & a Nick under the same.

" Samuel Elwell Jr Ear mark is a Crop of the left and a hole in the rite "

The first instance we have discovered in the records of the towns of this county, of a master manumitting his slaves, occurs, to its praise be it said, in this town by Samuel Field; and what gives it more importance, is the fact that our fathers had just entered on the struggle with the Mother Country, for freedom from foreign domination.

But there is a drawback to this noble instance of magnanimity, found on the records as late as 1826, in the sale of the town-poor. Slavery is a misfortune, and poverty is not necessarily a crime; and for governments to treat it as such, would only be making the rich richer, and the poor poorer.

"To all persons, unto whomsoever these Presents shall come greeting. Know ye, that I Samuel Field of Oblong in the County of Dutchess and Province of New York, For and in consideration of the free rights and liberties of all mankind, and conceiving it unlawful for a Christian to hold any of his fellow creatures in bondage for term of life: Do hereby, from and after the thirteenth day of the fifth month, Called May which shall happen in the year of our Lord 1780 give unto my Mulatto Man, bred by me, Known by the name of Phillip, his full freedom, to act & do in business for himself as of his own proper right as a free Man—And to be free from all manner of claim or command in any Kind of service whatsoever, either by me my heirs, executors, administrators or assigns for ever. In witness whereof, I have hereunto set my hand & seal this tenth day of the second Month called february in the year of our Lord 1776.

"SAMUEL FIELD (L. S.)"

Signed & sealed in
ye presence of us
PETER FIELD
ABEL CLOSE.

"Sale of the Poor made April 22 1826

	Nancy Binnit	to James Hains	$25,37
	Abigah Crane	" Henry Mead	33,50
	George Dudley	" Abner Gay	35,80
	Ebenezer Wixon	" Chancey Higgins	15,00
	Birch & wife	" Henry Cole	79,94
June 1	Esther Lawrance	" James Hains	23,87
July 1	Joseph Leonard	" James Hains	34,97 "

"At a Town Meeting held the 6th day of April 1779 it was Voted that Jabez Elwell make a Pair of Stocks upon the Cost of this precinct.'"

Sodom Corners.—A small village on the East Branch of the Croton River, in the centre of the town, about six miles East of the Court-house. Barber, in his Historical Collections, calls it Hatsville, which name must have been given to him by some one who desired that the name of Sodom should be dropped. We do not wonder, however, that any of its present peaceable inhabitants should desire to get rid of a name, which has connected with it such odious associations.

When we approached this quiet little hamlet, we looked around us for Gomorrah, but failed to discover it. And when fairly quartered in its centre, we smelled no brimstone, but saw fire; but that was where it should be—in the stove. It was named "*Sodom*" by way of reproach, in consequence of the unusually wild and wayward character of some of its *B'hoys*, in days long gone by. But for peace and quietness, sobriety, and industry, so far as we could discover, it is not now excelled by any other village in the county. A Post Office is located here.

Milltown.—A small village about two miles east of Sodom Corners. It is so called from a mill located there. A Post Office is also established here.

Doansburgh.—A few houses located in the northerly part of the town, in the neighborhood of the first Presbyterian church. This church occupies the site of the old one, in which the Rev. Elisha Kent, grandfather of the late Chancellor Kent, first preached, on his arrival in this town, after leaving the Presbyterian

Church at Newtown, in the State of Connecticut, in 1740. It subsequently "became known as Kent's Parish." Chancellor Kent was born at this place, in an old house which stood about three rods distant from the Post Office; and which was taken down about twenty years since. The east sill of the store-building at this place, now covers the spot, which, in the Chancellor's boyhood, was covered by a rock as high as a man's head, of a pyramidal form, with artificial steps in the side of it from its base to the top. The Hon. Reuben D. Barnum, Clerk of Putnam county, some twenty-five years ago, blasted the rock into pieces. Shortly afterwards, the Chancellor visited the place of his birth—the scene of his nativity—to commune with the hallowed recollections of the past, and the golden memories of childhood's sinless days. He expressed his regret at its demolition, as it had been a source of great pleasure and amusement to him in his boyhood, to climb to its apex, and indulge in those day-dreams which characterize that sunny period of existence. He also visited the house, and ascended to the chamber where he was born. He seemed excited with all the rapturous feelings of boyhood, and exclaimed, "*Here is the room where I was born—the chamber where my existence commenced.*" This place is named in honor of Benjamin Doane, who lived here.

Joe's Hill.—A beautiful and romantic eminence in the east part of the town, the west end of which terminates on the east bank of the east branch of the Croton River. A part of this Hill belonged to David Paddock, deceased, who came from Cape Cod, in the State of Massachusetts, when seven years old, with

25*

his parents. The same part is now owned by his son, David B. Paddock.

About seventy years ago, it was rumored that there was a silver mine on the north side of this *Hill*. Marvellous stories were told concerning the manner of its discovery; and that it was a charmed spot, to which no man would be permitted to approach, without having first taken an oath of secrecy, and become imbued or subject to the mystic *spell* that rested over the *Potosi* of the town of Southeast.

About sixty years ago, the excitement was at its height. It drew within its vortex some men who were possessed of an easy credulity, and others who, from a knowledge of the character of their neighbors, relied upon their statements, and trusted in their reports. It having been noised abroad, that there was a silver mine of incalculable wealth here, two or three men from abroad, supposed, at the time, to be theoretical or practical miners, visited the hill and took up their residence near it. Two residents, Nathan Hall and Jehu Miner, also became believers in the existence of the mine. They, with the others who came from abroad, were called "*Pigeon Men*," by their neighbors.

Hall pretended to know the precise locality of the mine, and, when questioned by his friends and neighbors, urged the necessity of first subscribing to a secret oath. David Paddock, deceased, often conversed with Hall about the *mine*, but could get nothing satisfactory from him. Necromancy, divination, and mystic charms, formed the subjects of Hall's conversation when plied with direct questions touching the location of the hidden treasure. While young he went to Demarara, in English Guiana, in South America. In a few years he returned, alleging that he knew everything that had happened in his absence. The existence of the *mine* appears to have been a delusion, which seems to have increased with his age; and, as a ruling passion, was strongly developed on his death-bed. In this respect he was what is now termed a monomaniac—deranged in a single faculty of his mind, or in regard to a particular subject.

The elder Paddock asked him what use was it to a man to believe what he was destined never to know; and how could a man believe in what everybody else, himself, and a few others excepted, rejected as a delusion? Hall told him, that if he could be led along to a certain spot and see a stone move, which covered steps leading to the *mine*, he would then believe. "Yes," replied Paddock, "if I saw it with my eyes, and descended into the mine, I should believe; but I do not believe I shall ever see all of that, or that you have either."

Laboring under this delusion, which seemed ever uppermost in his mind, he induced his wife to believe, that in a short time there would be more silver money in Southeast than there ever had been in all preceding time. Mrs. Richards, now living near this Hill, asked her what had become of the large amount of silver which her husband had predicted would be as plenty as berries. She replied that "Nathan had been revealing something about the *mine* that he ought not to have disclosed, and *the mysterious spell* had moved over it."

Hall's brother told him, that when he came to die he wished to be present; and desired that he might be sent for in time to witness his departure. He was sent for, accordingly, but those who caused him to be informed that his brother's departure was near at hand, were not aware of the additional and secret reason, that had prompted Nathan's brother to desire that he might be sent for when Nathan was about to quit this world. The brother came, but Hall was so far gone as to be unable to speak, yet conscious of everything around him. His brother then told him, if there was a silver mine in Joe's Hill, and he knew it, to squeeze his hand, which he then gave him. Hall then gave it, as far as his strength permitted, a hard gripe; thus ratifying on his death-bed the belief he had taught and entertained while living. Nathan Hall died in this town about ten years ago.

Tone's Pond.—This handsome body of water lies in the westerly part of the town, is one mile long, half a mile wide; and is named in honor of an old negro of the name of *Tone*, who settled beside it. He was the slave of John Warring, deceased, enlisted and served in the Revolutionary war, on condition that he should have his freedom at its close. Having received his freedom after peace was declared, he married a woman half Indian and half negro; settled down beside this pond which soon became a great resort for fishing sportsmen. He kept boats for their accommodation, and furnished them with whatever was called for in the form of victuals and drink; a sort of fisherman's tavern, where everything appertaining to the sport is to be had at the shortest notice, for which liberal payment is made, and no credit expected by either party. He left quite a numerous and respectable number of descendants. It is said that one of his grandsons married a beautiful young white girl who shortly afterwards induced him to go south with her, where she sold him as a slave.

Little Pond is about half the size of Tone's Pond, and about half a mile west of the Presbyterian Church in the northerly part of the town.

Peach Pond.—This is a large sheet of water, the greater part of which lies in Westchester County. It is two miles long and one broad. The dividing line of the two counties runs through the north part of it, leaving a small portion of it in Putnam.

Corner Pond Brook.—This stream is so called in consequence of four towns cornering on it, viz.:— Danbury, Ridgefield, New Fairfield in Connecticut, and this town in New York. The north-east corner of this town is on it.

Daly Brook.—This stream empties into the Croton near Milltown, and about one mile north of the west end of Joe's Hill.

"Return of Militia Officers for South-east Precinct, Dutchess County, New York:

"Dutchess County, South-east Precinct Committee, August 21, 1775.

"Pursuant to a Resolution of Provincial Congress,

"Ordered that *Thomas Baldwin*, Esquire, and Mr. *Nathaniel Foster*, two of the Members of this Committee, notify the Militia of this Precinct, consisting of one Beat (lately commanded by John Field, as Captain), to appear on the 25th instant at the usual place of parade, that the said Militia, under the direction and inspection of the said *Baldwin* and *Foster*, may arrange themselves into a military Company, agreeable to said Resolution of Congress. That said Militia do then and there make choice of military officers by a majority of votes, to take the command of said Company; and that the said *Baldwin* and

Foster make return of their doings to the Chairman of this Committee.

" JOSEPH CRANE, Chairman."

"Having duly executed the above Order of Committee, we hereby certify that the Company of Militia of said South-east Precinct, agreeable to said Order, did assemble; and they have, by a fair majority of votes, made choice of Commissioned Officers to take the command of said Company, agreeable to the aforesaid Resolution of Congress, as follows, viz.: *William Mott*, Captain; *Benjamin Higgins*, First Lieutenant; *Ebenezer Gage*, Second Lieutenant; *Nathaniel Green*, Jun., Ensign.

"Test. "THOMAS BALDWIN,
"NATHANIEL FOSTER,
"A true copy of the Return.

" JOSEPH CRANE, Chairman."

"Return of Minute-officers in South-east Precinct, Dutchess County, New York:

"Dutchess County. South-east Precinct Committee, September 22, 1775.

"Ordered, That those persons who have arranged themselves in the character of Minute-Men in this precinct, do assemble themselves on the 26th instant, in order to choose out of their Company the several officers which, agreeable to directions of our Congress, are to command such Companies; and that *Thomas Baldwin*, Esqr., and Mr. *Nathaniel Foster*, Members of this Committee, do attend and inspect said choice, and make return thereof to the Chairman of this Committee.

"JOSEPH CRANE, Chairman."

"South-east Precinct, Sept. 26, 1775.

"We hereby certify that agreeable to the foregoing order, the Company of Minute-Men therein referred to did, on the 26th instant, assemble, and, under our inspection, made choice of *Joshua Barnum*, Jun., as Captain; *William Marsh*, First Lieut.; *Eliaakim Barnum*, Second Lieut.; and *Jonathan Crane*, Ensign

"THOMAS BALDWIN,
"NATHANIEL FOSTER."

The following is a letter from Joseph Crane, Chairman of the South-east Precinct Committee, to Egbert Benson:

"South-east Precinct, May 22. 1777.

"Dear Sir—Yesterday I saw one Allaby, a sergeant of Captain Dellman, taken prisoner at Ward's, with Major Dain; he made his escape from the guard-house in New York on the evening of the 15th inst.; gives a favourable account of the state of the prisoners taken with him, our worthy friend, Major Dain, excepted, who, he says, has been in close confinement during the whole of his captive state; that the only reason assigned therefor is, his having no commission with him; says he has often heard the Major lamenting the hardships to which he is subjected on that account, wishing for an opportunity to inform his friends thereof; that in addition to a train of painful circumstances consequent on a state of close confinement, the means of subsistence was rendered much more expensive. His supplies from home, I understand, have been much short of what I have before understood, twenty-five dollars being the full amount of what he has received; he has had the small-pox by inoculation pretty severely; is now in a good state of health. I presume a simple relation of facts will be sufficient to engage your attention, and that the earliest opportunity will be embraced for the relief of this worthy officer; anything in my power to forward the same will be with pleasure complied with.

"I have the pleasure to hear my son is well; has for some time been admitted to his parole, either to remain in New York or go to Long Island; has chosen the former; taken his board with Mr. Thomas Arden; had the small-pox favourably, and in all respects is as happy as a state of captivity admits of. Alleby says the enemy lost fourteen men in the action at Ward's; six of those they carried off wounded died between Ward's and Williams'; the seventh died as soon as they had got to Valentine's: that every commissioned officer, save one ensign, was killed; that on their arrival at King's Bridge, the commanding officer of that post came out of the door of his lodgings, when the prisoners were paraded, and says, ' Well, you have got a parcel of the d——d rebels, have you?' 'Yes,' replied the surviving ensign, 'but we have paid d——d dear for them; I am the only officer left alive!' to which there was no reply made.

"He further tells me, that the day before he left New York, he read in the papers an account of the enemy's loss in Danbury tour, estimated at between 3 and 400 men; that he often heard them say to one another, that the Danbury rout had been more expensive to them, in proportion to the number of their troops, than the Lexington tour. May Heaven grant that their cursed enterprises may still prove more and more expensive to them, till they become quite bankrupts.

"We are now in this quarter (as it were) lying on our arms, every hour expecting fresh visits from the Tryonites. A number of the enemy's ships are in the Sound. Yesterday morning upwards of twenty of them drew up against Fairfield, and appeared to be in a landing posture; the alarm reached us by 12 o'clock the same day; by night we were informed they soon came to sail again, and went westward of Norwalk. They are no doubt meditating another of their felonious enterprises, and I have the pleasure to assure you our people are evidently better disposed, as well as better prepared otherwise, to bid them welcome, than ever we were before: and the general say is, that in case Tryon is not gone to account for his former murders, 'tis hoped he will 'again grace his murderous train with his presence, and happily meet what Heaven has declared shall be the fate of him in whose skirts is found the blood of men.'

"You will overlook the blunders in this hurried scrawl, and accept the humble regard of, Sir,

"Your most obedient,
"JOSEPH CRANE.

"To EGBERT BENSON, Esqr."

"*Letter from Thaddeus Crane.*

"Kingston, August 9th, 1777.

Gentlemen—It being my misfortune, in repulsing the enemy at Ridgfield, on the 26th of April last, to meet with a wound from them, which confined me to my bed for a long time; and I was at great expense by loss of time and cost of doctor. I desire to know from your Honourable Council, whether I am to receive

any wages or relief from the State, and where to apply to get the same, if any is allowed. These from your honours' humble servant,
"THADDEUS CRANE.
"To the Honourable the Council of Safety of the State of New York."

JAMES KENT.*

James Kent was born the 31st July, 1763, in that part of Dutchess County, then called the Precinct of Fredericksburgh, now in the County of Putnam, in the State of New York.

His grandfather, the Rev. Elisha Kent, a native of Suffield, in the State of Connecticut, married the daughter of the Rev. Joseph Moss, of Derby, and was for some time a minister of the Presbyterian Church at Newtown, in that State.

He removed, as early as 1740, to the south-east part of Dutchess County, then wild and uncultivated, but which gradually increased in population, and became known as Kent's Parish.

He continued to reside there until his death, in July, 1776, at the age of seventy-two. His eldest son, Moss Kent, who, as well as his father, was a graduate of Yale College, commenced the study of the law under Lieutenant-Governor Fitch, at Norwalk, in Connecticut, and was admitted to the bar, in Dutchess County, in 1756. In 1760 he married the eldest daughter of Dr. Uriah Rogers, a physician at Norwalk, by whom he had three children, who are now living: James, the subject of this memoir; Moss, who was a member of the Senate of New York for four years, afterwards a member of Con-

* National Portrait Gallery.

gress, and first Judge of the Court of Common Pleas of Jefferson County, which office he resigned on being appointed Register of the Court of Chancery in 1817; and Hannah, who married William Pitt Platt, of Plattsburg. They lost their mother in 1770, and their father died in 1794, at the age of sixty-one.

When five years old, James, the eldest son, was placed at an English school at Norwalk, and lived in the family of his maternal grandfather until 1772, when he went to reside with an uncle at Pawlings, in Dutchess County, where he acquired the first rudiments of latin. In May, 1773, he was sent to a latin school at Danbury, in Connecticut, under the charge of the Rev. Ebenezer Baldwin, a highly respectable Presbyterian minister. After the death of Mr. Baldwin, in October, 1776, he was under different instructors, at Danbury, Stratford, and Newtown, until he entered Yale College, in New Haven, in September, 1777. At these different schools he was remarked as possessing a lively disposition, great quickness of parts, a spirit of emulation, and love of learning.

The pious Puritans among whom he lived were sober, frugal, and industrious; and the strict and soberly habits of those around him had their influence in forming his own. From their example, and the impressions received at that early age, he acquired that simplicity of character and purity of morals which he ever afterwards preserved, without losing his natural vivacity and playfulness of temper.

He has often mentioned the delight he experienced on his periodical returns from school to his home, in rambling with his brother among the wild scenery of his native hills and valleys.

The associations then formed rendered him an enthusiastic admirer of the beauties of nature; and, in after-life, during the intervals of business, he made excursions into every part of his native State, through New England, and along the borders of Canada, visiting each mountain, lake, and cascade; and while gratifying his taste for simple pleasures, preserving and invigorating his health.

In July, 1779, in consequence of the invasion of New Haven by the British troops, the college was broken up, and the students for a time dispersed. During his exile, having met with a copy of Blackstone's Commentaries, he read the work of that elegant writer with great eagerness and pleasure, and it so excited his admiration that he determined at the age of sixteen to be a lawyer.

He left college, after taking the degree of Bachelor, in September, 1781, with high reputation; and after passing a few weeks at Fairfield, to which place his father had removed on his second marriage, he went to Poughkeepsie, and commenced the study of the law, under the direction of Egbert Benson, then Attorney-General of the State of New York, and afterterwards one of the Judges of the Supreme Court.

His strong and decided attachment to jurisprudence could not fail to ensure his success. Besides, the books of English Common Law, he read the large works of Grotius and Puffendorf, making copious extracts from them, and, as a relaxation, perusing the best writers in English literature, of which his favorite portions were history, poetry, geography, voyages, and travels. He was temperate in all his habits, a water-drinker, and entered into no dissipation, not even joining in the ordinary fashionable amusements of others of the same age.

He was very far, however, from being grave, reserve, or austere; but was uniformly cheerful, lively, and communicative.

The love of reading had become his ruling passion, and when he felt the want of amusement, "he better knew great nature's charms to prize," and sought it in rural walks, amidst objects that purify and elevate the imagination. In September, 1784, he took the degree of Master of Arts at Yale College, and in January, 1785, was admitted an attorney of the Supreme Court. He went to Fredericksburgh with the intention of commencing the practice of his profession there; but the solitude of that retired spot soon became insupportable, and in less than two months he returned to Poughkeepsie, where, in April, 1785, he married Miss —— Bailey, a lady a few years younger than himself, and with whom he has since lived in the uninterrupted enjoyment of domestic felicity. He possessed at this time little or no property, but living with great simplicity in a country village his wants were few, and supplied at little expense. Young, ardent, and active, he felt no anxiety for the future; but engaged with increased alacrity in professional business and literary pursuits, so as to leave no portion of his time unemployed.

In 1787, he resolved to renew and extend his acquaintance with the Greek and Roman classics, which he had entirely neglected after leaving college. When it is considered that the only Greek book, at that time, read by the classes, in that seat of learning, was the Greek Testament, and the only Latin works, Virgil, the select orations of Cicero, and some parts of Horace. we may easily imagine how imperfect must have been that part of his education, the defects of which he was determined to supply. He began a course of self-instruction, with an energy and perseverance that mark a strong and generous mind. That he might lose no time, and pursue his various studies with method and success, he divided the hours not given to rest into five portions: rising early and reading Latin until eight, Greek until ten, devoting the rest of the forenoon to law; in the afternoon, two hours were applied to French, and the rest of the day to English authors. This division and employment of his time were continued with little variation, until he became a Judge. By this practice, he was under no necessity of encroaching on those hours best appropriated to sleep, and preserved his health unimpaired. If his mind became weary in one department of study, he found relief in passing to another; "from grave to gay, from lively to severe." He read Homer, Xenophon, and Demosthenes with delight. Though he afterwards relinquished the pursuit of Grecian literature, he continued to read the best Latin and French authors, and many of the former more than once. As large public libraries, if any then existed, were not within his reach, he began a collection of books which he has gradually increased to several thousand volumes; and he has often said that, next to his family, his library had been to him the greatest source of enjoyment. It fed, while it increased his appetite for useful knowledge, and cherished that love of literature that had grown and strengthened with his growing years.

In April, 1787, he was admitted a Counsellor in the

Supreme Court. He soon entered with ardor into the discussion of the great political questions which then absorbed the attention and agitated the minds of all. He could not long remain neutral between the two contending parties, and after a careful examination of the arguments of each, he, from the purest motives and with the clearest conviction, joined the federal side. He soon became the friend of Jay, Hamilton, and other eminent men of that party, with whom he uniformly acted, and to whose principles he has steadily adhered to the present day.

In April, 1790, he was elected a member of the State Legislature for Dutchess county; and again in 1792. In the Session, held in the city of New York, he took a zealous and distinguished part in the memorable question which arose in that body, on the conduct of the canvassers of votes given in the warmly contested election for Governor, in destroying those returned from Otsego county, by which means Mr. Clinton obtained a small majority over Mr. Jay, (then Chief Justice of the United States), who was the federal candidate. His writings, on that occasion, attracted much attention, and he became favorably known in the city. He was, at that time, nominated as a candidate for Congress, in Dutchess county, but his competitor, who adhered to the opposite party, succeeded by a small majority. During his attendance in the legislature, his principles and conduct were so highly respected, that he was urged by his friends to remove to the city, where he might find greater scope for the exercise of his talents, and more lucrative business in his profession.

He accordingly removed to New York, in April,

1793. The first month of his residence in the city was embittered by the loss of an only child, and for a time his prospects were clouded with sorrow. In December, he was appointed Professor of Law in Columbia College, and commenced the delivery of lectures, in November, 1794. The course was attended by many respectable members of the bar, and a large class of students. In the following winter, he read a second course; but the number of his hearers having diminished, he was discouraged from delivering another. The three preliminary lectures were afterwards published, but the sale of them did not reimburse the expense of publication. The trustees of the College conferred upon him the degree of Doctor of Laws, and he has since received similar honors from Harvard University and Dartmouth College.

In February, 1796, he was appointed a Master in Chancery, and there being, at that time, but one other, the office was lucrative. In the same year, he was elected a member of the Legislature from the city of New York. He delivered an address before the society for the Promotion of Agriculture, Arts, and Manufactures, at their anniversary meeting in New York, on the 8th November, 1796, which is inserted in the first volume of the transactions of the Society. It contains a rapid and animating sketch of the great natural and political advantages of the United States, and especially of the State of New York, for the advancement of the great objects of the Society, and the progress of the country, since that time, has more than realized the most glowing anticipations of its patriotic founders.

In March, 1797, he was, without solicitation and quite unexpectedly to himself, appointed Recorder of the city. This being a judicial office, was the more acceptable as well as more honorable; and being allowed to retain that of Master, the duties of both were so great, and the emoluments so considerable, that he gradually relinquished the more active business of his profession, to which he was not strongly attached. From constitutional diffidence, or habits of study, he appeared not to feel confident in the possession of the powers requisite to ensure pre-eminence as an advocate at the bar.

In 1798, Governor Jay, who knew his worth and highly respected his character, offered him the office of Junior Judge of the Supreme Court, then vacant, which he accepted. This appointment gratified his highest ambition. It placed him in a situation where he could more fully display his attainments, and have a wider field for the investigation of legal science. In accepting the office, he relinquished, for a limited income, all the flattering prospects of increasing wealth that had opened to him during five years' residence in the city. Though most of his friends doubted the wisdom of his choice, he never regretted it. And all who feel interested in the pure and enlightened administration of justice, have found reason to rejoice that he followed the dictates of his own judgment, in a matter so interesting to the honor and happiness of his after-life. On becoming a Judge, he returned to Poughkeepsie, but in the following year he returned to Albany, where he continued to reside until 1823.

When he took his seat on the bench of the Supreme Court, there were no reports of its decisions, nor any known or established precedents of its own, to guide or direct his judgment. The English law books were freely cited, and the adjudications of English courts regarded with the highest respect, and, in most cases, with the force of authority. The opinions of the judges were generally delivered orally, with little regularity, and often after much delay. The law was in a state of great and painful uncertainty. He began by preparing a written and argumentative opinion in every case of sufficient importance to become a precedent for the future. These opinions he was ready to deliver at the day when the judges met to consult on the decisions to be pronounced by the court. The other judges, pursuing a similar course, also gave their reasons in writing, supported by legal authorities. As he read with a pen in his hand, extracting, digesting, abridging, and making copious notes, the practice of writing opinions was easy and agreeable. Besides making himself master of all the English adjudications applicable to the points under examination, he frequently brought to his aid the body of the civil law, and the writings of eminent jurists of the countries in which that law prevails; especially, in the discussion of questions arising on personal contracts, or of commercial and maritime law, the principles of which have been so admirably unfolded and illustrated by Domat, Pothier, Valin, Emerigon, and others. Like Selden, Hale, and Mansfield, he thought law could not be well understood as a science, without seeking its grounds and reasons in the Roman law. From that great repository of "written wisdom," he drew largely, engrafting its sound and liberal principles on the hardy stock of the English common law. Thus commenced that series of judicial decisions which have enriched the jurisprudence of New York, and shed their influence on that of other States.

In 1800, he and Mr. Justice Radcliffe were appointed by the legislature, to revise the statutes of the State; and in January, 1802, was published their edition of them, comprised in two volumes octavo. Without venturing to change the phraseology of the laws, they confined themselves to the single object of placing together the various acts of the legislature relative to the same object, so as to bring the original enactments, and all subsequent additions and amendments, into one act; and by a full and accurate index, to facilitate a reference to them.

In July, 1804, he was appointed Chief-Justice of the Supreme Court, in which he continued to preside until 1814. We shall not here attempt to enter into any examination of the opinions delivered by him during the time he was a Judge of that Court. They are contained in sixteen volumes of Reports, from January, 1799, to February, 1814; and the judgment of the public has long since been formed on their merit and importance.

In February, 1814, he was appointed Chancellor. The powers and jurisdiction of the Court of Chancery were not clearly defined. There were no precedents of its decisions, (if we except what might be gleaned from a few cases heard in the Court of Errors, on appeal, and reported by Mr. Johnson,) to which reference could be made in case of doubt; and it is a fact, that during the whole period of his sitting in Chancery, from 1814 to 1823, not a single opinion or dictum of his predecessors was cited. Without any other guide, he felt at liberty to exercise such powers of the English Chancery, as he deemed applicable, under the constitution and laws of the State, subject to the correction of the Court of Errors, on appeal. As to the course of equity to be administered, it was to him, in effect, as if the Court had been then newly established. The causes before the Court were managed by a few lawyers. He opened wide its doors; and his kindness and affability, his known habits of business and prompitude of decision, attracted many to the Court. The number of causes rapidly increased, and it soon acquired the most strenuous and unceasing efforts of his active mind to hear and decide the cases brought before him. Besides his attendance during the regular terms of the Court, he was, at all times, easy of access at his chambers; so that no one ever complained of delay, as to the hearing or decision of his cause. He considered the causes in the order in which they were presented or argued, and did not leave one until he was fully prepared to deliver his judgment upon it. He read the pleadings and depositions with the greatest attention, carefully abstracting from them every material fact: and having become familiar with the merits of the cause, he was able, unless some technical or artificial rule was interposed, by his own clear moral perception, to discover where lay the equity of the case. Not content, however, with satisfying his conscience as to the justice of his decision, he was studious to demonstrate that his judgment was supported by the well-established principles of equity to be found in the decisions of the courts of that country from which our laws have been derived. His researches on every point were so full, as to leave little or nothing to be supplied by those who might afterwards wish to have his decisions re-examined, or to test the correctness of his conclusions.

Accustomed to take a large view of jurisprudence, and considering law not as a collection of arbitrary and disconnected rules, but rather as a science founded on general principles of justice and equity, to be applied to the actions of men in the diversified relations of civil society, he was not deterred, but animated, by the novelty and intricacy of a case; and while his mind was warmly engaged in the general subject, he sought, rather than avoided, difficult points, even when the discussion of them was not essential to the decision of the main question between the parties; so that nothing was suffered to pass without examination.

His judicial opinions are, therefore, uncommonly interesting and instructive to all, but especially to those who have commenced the study of the law, and aspire to eminence in that profession. The decisions in Chancery are contained in seven volumes of Reports.

On the 31st July, 1823, having attained the age of sixty years, the period limited by the Constitution for the tenure of his office, he retired from the Court, after hearing and deciding every case that had come before him. On this occasion the members of the bar residing in the city of New York, presented him an address, from which, as coming from those most competent, by their situation, to form a just estimate of his judicial character and services, we cannot refrain from giving some extracts. After speaking of the inestimable benefits conferred on the community by his judicial labors for five and twenty years, they observe: "During this long course of services, so useful and honorable, and which will form the most brilliant period in our judicial history, you have, by a series of decisions in law and equity, distinguished alike for practical wisdom, profound learning, deep research, and accurate discrimination, contributed to establish the fabric of our jurisprudence on those sound principles that have been sanctioned by the experience of mankind, and expounded by the venerable and enlightened sages of the law.

"Though others may hereafter enlarge and adorn the edifice whose deep and solid foundations were laid by the wise and patriotic framers of our government, in that common law which they claimed for the people as their noblest inheritance, your labors on this magnificent structure will for ever remain eminently conspicuous, commanding the applause of the present generation, and exciting the admiration and gratitude of future ages."

A similar address was presented to him by the members of the bar in Albany, and also by those from the different Counties of the State, attending the Supreme Court at Utica, in August following. In the latter it is observed, that, "In the space of little more than nine years, an entire and wonderful revolution in the administration of equity has been accomplished;" and a reference is aptly made to the account given by Sir William Blackstone of a similar revolution in the English Chancery by Sir Henage Finch, afterwards Earl of Nottingham, who became Chancellor in 1673. "The necessities of mankind," says that writer, "co-operated in his plan, and enabled him in the course of nine years, to build a system of jurisprudence and jurisdiction upon wide and rational foundations." In the same address, speaking of their intercourse with him as a Judge, they called to mind " so many instances of personal kindness—so many scenes of delightful instruction—so many evidences of pureness and singleness of heart—such a uniform and uninterrupted course of generous, candid, and polite treatment, that we are unable to express the fullness of our feelings, and can only say that our affection for you as a man, almost absorbs our veneration for you as a Judge."

In these addresses, the bar were led to express a doubt as to the wisdom of that clause in the political constitution of the State, which " compelled him in the full enjoyment of his intellectual faculties, to relinquish a station he had filled with such consummate ability." And, in this case, at least, the application of the policy of that provision might well induce them to call in question the wisdom and expediency of so singular a limitation.

In August, he visited the Eastern States, and on his return home, he became apprehensive that after being so many years actively engaged in discharging the duties of a public station, the sudden transition to privacy and seclusion might produce an unfavorable effect on his health and spirits. He soon determined to remove to the city of New York to open a law school, and to act as chamber counsel. The trustees of the College again offered him the professorship of law in that institution, which he accepted; and, in 1824, he prepared and delivered a series of law lectures, on a more comprehensive plan than that pursued in his former course. He also gave private instruction to students, who resorted to him from various parts of the United States. His parental kindness towards the young, and the frankness and affability of his manners, won their affection without diminishing their respect; and his conversation and example could not fail to inspire that ardor and emulation so conducive to their progress and success. His high reputation as a judge induced many, not only in the city, but in distant places, to consult him on difficult and important questions, and, instead of the brief answers usually returned by counsel, he gave full and argumentative opinions. Many causes actually pending in Court, were, by the agreement of the parties, submitted to his final decision. He had continued for some years thus usefully and agreeably occupied, when, having discontinued his law lectures, he began to revise and enlarge them for publication; and, in November, 1826, appeared the first volume of the "Commentaries on American Law." This volume includes three parts; the Law of Nations, the Government and Constitutional Jurisprudence of the United States; and the various sources of Municipal Law. The second volume was published in November, 1827, the third in 1828, and the fourth in 1830. The three last comprise the law concerning the rights of persons, and personal and real property.

He has treated the several subjects comprised under these extensive and most important titles—the rights of persons and the rights of property—in a manner more full and satisfactory than Blackstone; and has introduced many others, not found in the work of that author, with numerous references, quotations, and illustrations, the result of his various and extensive reading, highly pleasing and instructive to the student. He has left untouched the subjects of private wrongs, and the mode of pursuing their remedies by actions in courts of justice; of the powers and jurisdictions of judicial magistracy; and of public wrongs, or the law concerning crimes and punishments, which occupy the third and fourth volumes of the English Commentator.

The work of Sir William Blackstone, by the elegance of its style, its lucid arrangement, and finished execution, is so well adapted to render the study of the law attractive, and to give a knowledge of the constitution and laws of England, well deserving the attention of every liberal mind, that it has been (though, for many years, more from necessity than choice,) very properly placed in the hands of every student; but as much of those admirable commentaries relate to the political constitution of England, so different from our own—to its peculiar institutions, and to rights and duties, public and private, not existing in this country—an American work, exhibiting our own constitution, laws, institutions, usages, and civil relations, had been long wanted. In the full maturity of his understanding, with a mind long habituated to legal investigations and researches, and with sound and enlightened views of jurisprudence, no man, perhaps, could have been found better fitted than Chancellor Kent to execute such a work, and it may diminish, in some degree, the regret felt for the loss sustained by the public and the legal profession, in being deprived of his valuable services on the bench, to know how usefully to the world and honorably to himself, he has employed his time and talents in its performance.

The limits prescribed to this brief memoir will not permit us, if it were proper, to go farther, or to enter into a particular examination of the merits of this masterly work. The first edition of the Commentaries having been exhausted, he published a second in April, 1832, carefully revised and greatly enlarged. For one who has done so much for the improvement and diffusion of legal science, and who has now advanced to the limit ordinarily assigned to the duration of human life, it would be unreasonable to ask or expect more; but while he appears to feel none of the infirmities of age, or to seek indulgence or repose, we cannot suppress a wish, that he may yet be induced to present his view, also, of that system of equity and jurisprudence, to the formation and illustration of which his own judicial labors have so largely contributed.

Having been elected President of the New York Historical Society, he delivered, by request, a public discourse, at their anniversary meeting, on the 6th December, 1828.

In this elegant and instructing address, he very appropriately notices the principal events in the history of the Colony and State of New York, to the end of the Revolution, and mentions, with merited praise, some of the eminent patriots and statesmen of New York, who so ably assisted in achieving that Revolution, and in securing its blessings to their posterity. If our attention could be oftener drawn from the absorbing pursuits of wealth and ambition, or the contests of selfish demagogues, to the contemplation of such illustrious examples of wisdom and virtue, we might find more perfect models for our imitation, and haply feel our hearts warmed with that pure love of country which glowed in their breasts.

At the request of the Phi Beta Kappa Society, of Yale College, a literary association formed in 1780, of which he was an original member, and comprising the most distinguished graduates of that seminary, he delivered a public address, at the anniversary meeting of the associates, on the 13th September, 1831. This discourse, in which he takes a historical survey of the College, from its origin in the beginning of the last century, and sketches the characters of its pious and learned founders, supporters, and instructors, is replete with generous feelings and just sentiments on literature and education. Alluding, towards the close, to his own class, of whom twelve (out of the twenty-five) were then living, and most of those present; he makes this natural and striking reflection: "Star after star has fallen from its sphere. A few bright lights are still visible; but the constellation itself has become dim, and almost ceases to shed its radiance around me. What a severe lesson of mortality does such a retrospect teach! What a startling rebuke to human pride! How brief the drama! How insignificant the honors and 'fiery chase of ambition,' except as mental discipline for beings destined for immortality."

In the brief notice which we have taken of the principal events in the life of this eminent jurist, we have adverted to some of the distinctive qualities of his character; and it will be perceived how pure, virtuous, upright, and honorable, that life has been, the full delineation of which must be reserved for some future biographer. Though not passed in scenes that attract the general gaze of mankind, or exite the admiration and applause of the multitude, it has been highly distinguished, affording a bright and instructive example of industry and perseverance in the pursuit of useful knowledge, and of unwearied diligence in the discharge of every duty, public and private.

Chancellor Kent has three children—a son and two daughters; the former was admitted to the bar a few years since. Happy in his family—amiable, modest, and candid in his social intercourse—kind, indulgent, and affectionate in his feelings—it would be pleasing, if it were proper, at this time, to speak of him in those private relations which awaken the best affections and warmest sympathies of our nature. With a sound constitution, strengthened and preserved by temperance and moderate exercise, he has enjoyed that perfect and uninterrupted health which is rarely the lot of the studious and sedentary. Possessing a cheerful temper, and a lively consciousness of existence, that fits him for enjoyment, he seems to have experienced, in a high degree, those blessings for which the Roman poet bids the rational inquirer after happiness to supplicate heaven, and those gifts have not been wasted or misapplied:

> "Semita certe
> Tranquillæ per virtutem patet unica vitæ."

Since the foregoing was copied, death has removed this esteemed and venerable man from among us. At the time of his death, "his family consisted of his two daughters and an only son, the learned and well-known Judge of the First Circuit, William Kent, who resigned the office of Circuit Judge some years since, and more recently gave up his Professorship at Cambridge, that he might cheer the latter days of his venerated and excellent father by his company and personal attentions.

"Less than a year ago, Chancellor Kent was one of the pall-bearers of his friend *Timothy Dwight*, and was then as erect, hale, and active as a man of fifty. At 84, he was somewhat deaf, but his capacity for work was still wonderful, his conversation interesting and animated, and his temperament as vivacious as when he was thirty years younger. He was unwell but for a short time before his death, which took place at half-past eight o'clock, December the 12th, 1847, at his residence in Union Square.

"The Courts adjourned out of respect for his memory, and both Boards of the Common Council adopted resolutions in honor of his name and character.

He was an exemplary Christian, a steadfast friend, an affectionate father, a tender husband, an ardent patriot, and a true lover and defender of his country's rights. So highly are his works esteemed abroad, that the Lord Chief Justice of England, Baron Denman, wrote to Judge Kent, some years since, to acknowledge the indebtedness of the legal professson throughout the world to him for his able Commentaries.

"Chancellor Kent would not allow his Commentaries to be stereotyped, but kept watching the decisions of the tribunals of America, England, and other parts of Europe, in matters involving important legal principles, with which he enriched his favorite work from time to time."

Capt. Joshua Barnum.—The ancestors of this gentleman were from England. About 1650, three poor boys who were brothers, of the name of Barnum, emigrated from England to this country. They were unable to pay their passage-money, and sold themselves, for a limited time, to pay the sum advanced to the Captain of the vessel, by their purchasers. A man of the name of Canfield, of Connecticut, bought enough of the time of the brother who was the grandfather of the subject of this brief memoir, and great grandfather of the Hon. Reuben D. Barnum of Carmel, to reimburse him for the passage-money. Canfield proved to be a hard master, and treated the boy with a good deal of severity. The young fellow said but little, put up with the treatment he received, but his countenance revealed a steady purpose, and his eye seemed to say, " This is not to last always; and when it does end, what has been *sauce* for the goose shall be sauce for the gander." He served his time out faithfully, and on the morning of his discharge, Canfield called him into his room, and bade him take a seat beside him. Young Barnum, with eyes flashing at the recollection of past wrongs, accepted the proffered chair; and Canfield thus addressed him: " You are now about to leave me; I hope you'll do well in the world and remember the lessons I have taught you. Upon the whole, I must say you have been a good boy, excepting some little matters. I am satisfied; are you?" Barnum, to whom every word appeared like fresh outrage and insult, because they were given by one whom he knew cared about as much for his welfare as he did for the dog that barked in the kennel, seized him by the coat-collar, and drawing him across his lap, he gave him half a dozen "digs in the short ribs" and elsewhere, and, throwing him from him, said, " I am satisfied too; I hope you'll remember the lesson I've given you. Good bye, Sir!"

He took up his bundle and departed, again to buffet tyranny and battle with adversity. After meeting with some reverses, he hired himself to a respectable farmer, whose daughter, in a few years thereafter, he married and left issue. Of their history, we are not informed; and we now return to the subject of our sketch.

Capt. Barnum came from the town of Danbury in Connecticut, and settled where Elijah Barnum now lives, on the west side of the east branch of the Croton River. He had three sons, Stephen, Joshua, and Jonathan; the latter is the only one now living; and two daughters, Martha and Adah. Martha married Reuben Done; Adah married Jeremiah Gage; and both are dead. Stephen, the eldest son, was the father of the present Clerk of Putnam county. Capt. Barnum was in the battle at Ward's house in Westchester, was wounded, taken prisoner, marched to New York, and confined in one of the prison-ships, which were stationed at the Wallabout, near Brooklyn. There were several condemned hulks there, used for the confinement of American citizens, soldiers, and seamen, taken prisoners by the British. There were two hospital-ships, called the Hope and Falmouth, intended for the sick, anchored near each other, about 200 yards east of another named the "Old Jersey," which was the receiving ship. They were all, truly, ships of death, for in no other prisons, either on land or water, were so great an amount of suffering endured by civilized men. Into one of these he was thrust with a broken leg, to which but little attention was paid by the British. Inflammation, induced by the noxious vapors of a crowded vessel and his wounded leg, soon destroyed his right eye; and a small bunch of bones dropped from his broken limb, which are now in the possession of Doctor Barnum, of this town. While a prisoner, he and others were compelled to lie on the bare floor, with his leg undressed. A man by the name of John Roberts and six others died in one night, and were unremoved for the space of twenty-four hours by the British, although begged to do so by the whole body of prisoners.

During his confinement he became very feeble and debilitated, and suffered extremely from a violent headache. After he was released, he got a barber to shave his head, which relieved him. About four years before his death, he was again attacked with the same kind of headache; he insisted on having his grandson, Col. Barnum, to lather and shave his head, believing in the virtue of the previous operation. It was done, and the pain left him.

After the war he went to New York, where he had business to transact, and bought half a pound of Bohea tea, which he brought home to his wife. None, it seems, had been used at that time in this part of the country, and his wife was at a loss how to prepare it for use. Here was a dilemma not anticipated by the Capt., when he purchased the tea; and after resolves and counter-resolves by the family, it was resolved that the female counsellors of the neighborhood should be called together to decide the mooted point. They accordingly assembled, and, in high debate, proceeded to give their separate opinions respecting the mode in which this new *comer*, as yet only found in the circles of higher life, should be treated, and its virtues disclosed. One was for putting it in the pudding-bag and boiling it in milk; another was for frying it in a pan with a little butter and water; a third was for putting it in the dish-kettle and boiling it. This proposition determined the conclave at once in its favor; accordingly the half pound of tea was put in with a sufficiency of water, and duly boiled. For a teapot, they made use of a large earthen pitcher, and for tea-cups, bowls. Some drank more and some less; but the one who had recommended the dish-kettle, drank by far the largest quantity, alleging that she wanted to "*diskiver* its aristocratic qualities, if it had any." They went home, and the next morning the Capt.'s wife saw the one who had drank so freely of the tea, standing at her door, and asked her how she liked it. The reply was,—"*Last night I didn't sleep a wink—not one blessed wink! I might about as well have attempted to sleep with a thorn under me, as that ar tea inside—it's the plaguyest stuff I ever did drink.*" On inquiry it was found that all had been in the same predicament.

Captain Barnum was born in 1737, and died in October, 1823. He came to this town about 1755, and settled on lot No. 11, on the west side of the centre line of the Oblong. He was six feet high, strong and muscular, and capable of enduring great hardship and fatigue. During the Revolution he was absent a great deal from his farm and family.

His wife, in his absence, exercised supervision over the farm, and with the assistance of her sons, who were young, she managed to have it cultivated sufficiently for the wants of the family. At this time a young man of the name of Doty, whose family lived near Captain Barnum, and who had enlisted in the American army, deserted. His mother was very much alarmed, lest, if taken, he would be hung as a deserter.

In her trouble she applied to Mrs. Barnum, who, being a mother, could well appreciate the maternal feeling in others. By her direction, her oldest son, Stephen, the father of Col. Barnum, and then only sixteen years of age, took young Doty's place as a substitute, and served one campaign. Captain Barnum possessed a clear head and a strong mind, and was a patriot in principle and practice. His wife was equally a remarkable woman; and although she could sympathize with the innocent of her own sex, she was unwilling that the defenders of her country should number one less; and with all the heroic feeling of the Spartan mother of old, she gave her own beardless boy as a substitute to combat tyrants, and battle for the liberties of the country that gave her birth.

TOWN OF KENT.

As we have before stated, this town was a part of the "Fredericksburgh Precinct," which originally embraced the whole of the County. After the organization of the "*Philips' Precinct,*" which embraced nearly one-third of the west end of the County, the former contained this town, Carmel, Patterson, and Southeast.

By the Act of the 7th March, 1788, the term "*Precincts*" was dropped, and "*towns*" substituted; and an additional town organized, which was called Southeast.

By that *Act,* this town, including the now towns of Carmel and Patterson, was called FREDERICK'S TOWN.

About the time that Carmel and Patterson were taken from it, it was christened by the Legislature with its present name in honor of the Kent family, who were early settlers in this County, and greatly distinguished for their talents, intelligence, and manly virtues.

A large part of this town is rough and unproductive; the western part of which is covered by the central Highlands. There is, however, some excellent land, which is under a good state of cultivation. The eastern part is hilly, and well adapted to grazing, to which the farmers generally, we believe, give their attention. It is centrally distant from New York about 60, and from Albany about 101 miles. Farmers' Mills and Coles' Mills, are the only villages in it; the former in the north-west, and the latter in the south part of the town.

EARLY SETTLEMENT.

This town was settled by the Boyds, Smallys, Wixons, Farringtons, Burtons, Carters, Meritts, Barretts, Luddingtons, with a few others from Massachusetts and Westchester.

Zachariah Meritt settled in this town about 1750, and built a log-house in the meadow just east of the residence of Stillman Boyd, Esq. He planted himself, as it were, in the very midst of the Indians, who had a settlement there. We have been informed by the Hon. Judge Boyd, that he has ploughed up on the same meadow, more than two bushels of oyster-shells and arrow-heads.

During the Revolution, Meritt took the British side of the question, and his land was confiscated by the State.

About this time a family of the name of Jones settled in Peekskill Hollow, where John Barrett now resides.

The Boyds are of Scotch descent. The great-grandfather of this family in this town came from Scotland to New York City; and from thence to Westchester.

Ebenezer Boyd, grandfather of Bennet and Stillman Boyd, came from Westchester, and settled where Stillman Boyd now lives, about 1780. There were three brothers who emigrated from Scotland to this country during the "Rebellion" of the partizans of the Stewart dynasty in 1745. One of them settled at Albany, known as Gen. Boyd, and died at the advanced age of 114 years.

Another settled in the lower part of Westchester, and was great-grandfather of the family in this town. The other brother settled at New Windsor, in Orange County, and was the ancestor of the Boyd family in that County.

A man of the name of Joseph Farrington was about the first settler at Farmers' Mills. During the "hard winter," a man of the name of Burton, put up the first grist-mill at that place.

About 1760, James Smally, Reuben, Robert, and Pelick Wixon, came from Cape Cod to this town, and settled about a mile east of Stillman Boyd. They were of English descent. A family of the name of Cole were also early settlers, as we are informed; but are unable to say at what time they came.

Col. Henry Luddington, or " *Luddinton,*" as it was formerly spelt, made a settlement in this town about 1760. He was born in Connecticut; but his father emigrated from England. He settled, where his son still resides, in the north-west part of the town, known in the Revolution as " *Luddinton's Mills."* A man by the name of Carter had settled at the same place a few years before.

Col. Luddington was one of the most active, energetic, and unflinching patriots that was found in this part of the country during the Revolution; and much do we regret our inability, from the want of materials, to do justice to the character and sterling virtues of this Revolutionary patriot. The governmental records, however, show him, in connexion with the Barnums, Cranes, and a few others, to have been one of the bold defenders of our country's rights. He left six sons and six daughters; of whom only two of the former and one of the latter are now living.

Extract from "Fredericksburgh Records A."

"April ye 7th Day and first Tuesday 1747.
Matthew Roe, Clark
Supervisor Chosen Samuel Field

Constables Chosen Viz:

Joseph jacocks   John Dickeson
George Huson   William Bruster

    Nathan Taylor Senr   Colloctor
    Joseph Lane   Seesser
    Capt: James Dickeson   Seessor

High Way Masters Chosen viz:

Jacob Vandweel   James Seers
Joseph Husted   Joseph Crane
Richard Curry   Samuel Field
Isaac Rhoades   Daniel Townsend
George Curry   Uriah Townsend
William Gee   ——— Barttlett
William Sweett   Caleb Heason

Pounders chosen viz

Thomas Kirkkun   John Gee and   Amos Dickeson

Fence Viewers chosen viz

Daniel Townsend   Isaac Roads
Amos Dickeson   Isaac three hill
Abraham Smith   John Rogers

The following named persons appear to have been in the Fredericksburgh Precinct in 1747, who were freeholders, or occupying land as tenants. We first remark, however, that this precinct included, at this time, all of the towns now embraced in Putnam

County, viz.: Kent, Patterson, Carmel, Southeast, Putnam Valley, and Philipstown. In 1772, Philips' Precinct, embracing both Philipstown and Putnam Valley, was erected; and, in 1773, the Southeast Precinct was organized. By the Act of March 7th, 1788, "for dividing the counties of this State into towns," the Precincts were changed to towns; and hence Philips' Precinct became Philipstown; Southeast Precinct, Southeast town; while the remaining part of Fredericksburgh Precinct was called Frederick's town, embraced the new towns of Kent, Patterson, and Carmel. Barber, in his Historical Collections, says, that Southeast was organized in 1795; but this is evidently a historical mistake, as the Act, above alluded to, shows.

Cristoph fowler,
William Gee,
William Taylor,
Thomas Kirkun,
John Drake,
Rickcobus Cartwright,
Samuel Field,
John Ryder,
Jeremiah Calkin,
John Moberry,
George huson,
Isaac Roads,
Benjamin Brundage,
Vallentine Perkins,
Dan'l Townsend,
David Paddock,
Uriah Townsend,
Will Hunt,
Old Cole,
Abraham Smith,

Joseph Lane,
Amos Dickinson,
James Kirkun,
James M. Creedy,
Uriah Hill,
George Curry,
Edward Ganong,
Richard peters,
John Williams,
Andrew Roble,
Ephraim Smith,
William Smith,
William Drake,
Zedekiah Kirkun,
Thomas Kirkun,
George Scott,
Isaac Rhoads,
John Names,
Joshua Hamblin,
Joseph Hopkins,

Abraham Lock,
Caleb Heaz,
Michell Sloat,
Elijah tompkins,
Daniel Parish,
Edward Gray,
John Dickinson,
Absalom Smith,
Sam'll Hunt,
Jonathan Lane,
Jonathan O'Brien,
Isaach Horton,
Richard Cory,
Thomas Devenport,
Joseph Mead,
Thomas Paddock,
Gabriel Knap,
Joseph Chatoren,
Bartlit Crondy,
John Backer,
Israel Smith,
Thomas Townsend,
Benjamin Jacox,
Nehemiah Horton,
Cornoloes Tompkings,
Elisha Cangs
James Akely,
David Smith,
Isah Jacox,
Isaac treehill,
Cristefor Alley,
Moses Dusenberry,
Jeremiah Jiffers,
Isaac finch,
Andres Barger,
George Feelds,
Amos fooler,
francis Beacker,
Robert fooler,

Isaac Merrick,
David Merick,
John Harick,
John Thorne,
Schirran Travis,
Joshua Parrish,
Jacob Parrish,
John Ganong,
Joseph Dewey,
Noah Burbank,
David Sturdeuant,
Thomas Crosby,
Pelig Baley,
Isaac Barton,
Thomas Colwell,
Matthes Burgus,
Israel Taylor,
James Bell,
William Stone,
Edward Rice,
John Sprag,
Isaac Smith,
Richard Rodes,
Thomas Philips,
Jonathan Briant,
John tarbe,
William Deusenberry,
Solomon Jenkins,
Josiah Forgason,
Dan'l Crawford,
stephen faronton,
John Landon,
John Mead,
John Meeks,
Gilbert Travis,
Joseph Stateer,
Nehemiah Wood,
John Heaus,
David Sears,

John Larnce,	Joseph Colwell,
Joseph Ganong,	Jacob Ellis,
Elisha Cole,	Regam Parrish,
Elnathan Gregry,	Bethnell Balleau,
William Sturdeuant,	John Addams,
Nathan Birdsall,	Ase Parrish,
Jabez Berry,	Nathaniel Nickerson,
Benamin Jackish,	Zenis Nickeson.
Matthies Burns,	

There were others, doubtless, who resided in this Precinct, whose names are not found on the town book. Beside other sources of information, we are satisfied from the names given above, that Fredericksburgh Precinct, at this time, embraced the whole of Philipstown under its original organization.

In the above extract from the Precinct Book, we find the names of Thomas Devenport, who, at that time, resided at Cold Spring, on the Hudson; and John Rogers, who kept a tavern a little south of Justus Nelson's mill, on the old post-road.

In the above extract of names, we have written them as spelled in the Record, using the same kind of letters, as are there found in the commencement of the christian and surnames.

Farmers' Mills.—A hamlet eight miles from Carmel, on the road leading from Cold Spring to Patterson. It contains 2 or 3 stores, 2 taverns, a Post-office, a grist, saw, fulling-mill, and a tannery. Its name is owing to the fact, that the first mill was erected for the accommodation of farmers in that vicinity; no sale-work being done there at that time for other parts of the country.

Coles' Mill.—A small collection of houses on the road from Cold Spring to the village of Carmel, about three miles west of it. It contains a grist mill, saw mill, and fulling mill. It is named after the family of Coles who settled the place. One of the west branches of the Croton runs through it.

Dick Town.—A small district of country lying south of the Cold Spring turnpike, near Justice Forshay's. A large number of persons of the name of *Richard* resided there. The nickname of Richard is *Dick*, by which they were generally called, and owing to the multiplicity of the name, their neighbors called it " *Dick Town.*"

Smalley Hill.—An eminence about half a mile east of Stillman Boyd's, and is named after the Smalley family.

White Pond.—A beautiful sheet of water about half a mile east of Farmers' Mills. It is nearly a mile long and half a mile broad, and named after a family by the name of White.

Forge Pond.—This pond is about sixty rods east of Farmers' Mills. Forty years ago, a forge was erected at the west end of it at its out-let; and hence its name. It was formerly a marsh. The White Pond runs into it. It is about one-fourth of a mile long and fifty rods wide.

China Pond.—A handsome body of water, nearly circular, one mile south of Farmers' Mills; and nearly a half a mile in length. The reason for its being so called, we are unable to give, but believe it was owing to the fact, that a small basket full of *China*-ware was thrown into it by a wife, to spite her drunken husband.

Pine Pond is about one and a half miles south of White Pond, three-quarters of a mile long and half a mile wide. Its margin was covered with pine timber, and hence its name.

Dean Pond is about one mile south of Samuel Townsend's, and almost round, being about 60 rods in length and breadth. It takes its name from the family of Deans, who lived near it. The above named ponds contain perch, pickerel, with the more common kinds of fish; and are resorted to in the summer season by the amateurs of the fishing-rod.

TOWN OF PATTERSON.

THIS town, as we have before remarked, was originally named Franklin, when organized in 1795. The back short lot of Beverly Robinson embraced nearly its whole area. Its population in 1840 was 1,349, and in 1845, 1,289. Situated in the north-east part of the county, its farmers are further removed from Peekskill in Westchester, where the eastern section of the county have heretofore, we believe, done their "marketing," than those of Putnam Valley, Carmel, and South-East. This inconvenience will soon be obviated by the New York and Harlem Railroad, which passes through this town, and which will afford a speedy transportation of the farmer's produce to the great metropolis of the country. Its surface is hilly, and with few exceptions, the high grounds are cultivated and productive. Those parts which we visited seemed better adapted to grazing than for grain, although some with whom we have conversed pronounce it well-nigh equally suited to both. Every variety of agricultural product is raised, and the soil, generally, is as productive as any in the county.

The streams are not numerous; the east branch of the Croton furnishes the greatest amount of hydraulic power, and that, at certain seasons of the year, is small. There are but two ponds in the whole town. It is bounded on the north by the south line of Dutchess county; on the east by Connecticut; on the south by Southeast and Carmel; and on the west by the town of Kent.

We should think that the agricultural wealth of this town, in proportion to its size, is equal to that of any other in the county. The soil in the vicinity of Patterson village is loam and sand: in other parts loam and gravel predominate.

We made but a flying visit to this town, and that was to " The City." We confess we are poorly prepared to do it justice. Its distance from our abode, and the wearisome ride necessary to be taken across the central Highlands over a bad road, must be our apology for the brief and imperfect notice it receives at our hands.

EARLY SETTLEMENT.

This town was principally settled by Scotch families, or their descendants. Some few came here from Westchester and New York, but the greater number were from Massachusetts and Connecticut. A large number of families from Cape Cod came into this town, Southeast, and Carmel, about the same time.

William Hunt, the grandfather on his mother's side of Daniel Hains, Esq., came from Rhode Island and settled down about three miles north of Haviland's Corner, in 1745. A tavern is kept there now by a man of the name of Sill. He had three sons, Samuel, Daniel, and Stephen, and one daughter, Deborah.

His brother Daniel, came about two years afterwards, and located about a mile and a half from him. Stephen, the youngest son, was a lieutenant in the Continental army, and was engaged in the battle at

White Plains. After the war, he settled down along the Mohawk, and died there. The other two brothers were tories, and took the "King's side;" after the war, one went north, probably to Canada, and the other settled in the southern part of Ulster County.

Shortly after Hunt's arrival, two men by the name of Bobbin and Wilmot, settled at Patterson Village; the former was a blacksmith, and the latter a saddler. When the war broke out, they both went to New York, and joined the British Army.

About this time, Capt. Daniel Heecock, and the grandfather of James Towner, Esq., made a settlement in the vicinity of the village, but at what particular time we are not informed. Asa Hains, who had served three years in the French War, at its termination, came to this town and settled where Reed Aiken now lives, about a mile east of Haviland's Corner. Previous to which, however, he lived about two miles south of it. He was born in Rhode Island, and came from thence to Long Island where he enlisted in the English army; and was ordered to the north. He had five sons, Enoch, Charles, William, Archibald, and Daniel, of whom the two latter are now living, and five daughters, Lucy, Abigail, Deborah, Sarah, and Betsy. Two of whom, Abigail and Betsey, are still living.

About 1748, Daniel Close settled at Haviland's Corner. Where he came from, we have not been able to learn. About the same time the Jones and Crosbys came and settled in the southern part of the town. Roswell Wilcox, settled about a mile south of Patterson Village at an early day, but whether before or after the old French war, we are not informed.

A few years before the French war, Matthew Patterson, grandfather of James Patterson, Esq., came from Scotland to New York city, and at the age of eighteen enlisted as a captain of a company of artificers in the British army, under Gen. Abercrombie. After the war he went back to the city, and a few years thereafter he removed to where his grandson, the above-named James Patterson, Esq., now resides. He was a member of the State Legislature nine years in succession, and for several, a County Judge.

He had three sons, John, James, and Alexander, and four daughters, Martha, Jane, Susan, and Margaret. All are now dead! He was a member of the Legislature when Col. Beverly Robinson's land in this county was confiscated. Having voted for the measure, probably it accounts for his refusal to become a purchaser under the Act, from feelings of delicacy as a legislator. He purchased, however, from one who had derived his title from the State, 160 acres, on which his gentlemanly descendant now resides.

About the same time, one Capt. Kidd, who came from Scotland, settled between Patterson Village, and Haviland's Corner.

At the time Burgoyne was attempting to force his way down the Hudson, Washington moved three brigades into this town, where they were encamped, in order to reinforce Gates, had he been forced to retreat, and check the enemy. They were encamped on the lands now owned by Judge Stone and Benjamin B. Haviland. An aged citizen asserts that one brigade was from Pennsylvania, one from South Carolina, and one from Georgia; the two latter we are inclined to think were from New Jersey or Connecticut, or from the latter and Massachusetts. Washington, with his life-guards, had his head-quarters where Legrand Hall now lives.

The few facts that we have hastily gleaned concerning the early settlement of this town, so hastily flung together, we have obtained from Daniel Hains, Esq., and James Patterson, Esq. Mr. Hains wants but a few days to being eighty-four years of age; and a more active, healthy, and sprightly old man of his age it has never been our fortune to encounter. His healthful and vigorous appearance did not more surprise us than the business we found him employed at. And what, reader, do you suppose it was? Some of you, perhaps, who suppose a man must necessarily be old at forty, and bed-ridden at fifty, will answer, "in the house, or in his bed;" you might, probably, long before you reach his age be found there. The simplicity that marked the age and mode of living, when he commenced life has long since disappeared, and luxury, with its untold evils in its train, is sapping the health and shortening the lives of the present generation. We found this old gentleman on a hill-side, so steep as to require some exertion to ascend it, with a crow-bar in his hand, engaged in raising or digging out stones and rocks. He told me that he was very near-sighted and somewhat deaf, but in all other respects felt perfectly well.

He is a remarkable man, and we can state a fact concerning him which probably cannot be said of any other living man in the county; and it is this: He has seen five generations of his own family—his grandfather, father, himself, his children, and grandchildren. His wife, who is still living, is about his own age, and when we saw her, she moved about the room unassisted by stick or cane with as much apparent ease as a woman of fifty. Alas, for the degeneracy of our times! but few of us can hope to reach the good old age already attained by this venerable couple. We live too fast, and more for show than for health and happiness. Judging from appearances, they have as good a chance for living ten or fifteen years longer as any of us.

EXTRACT FROM THE TOWN RECORD.

" At the first town meeting of the Freeholders and inhabitants of Franklin held at the House of James Phillips on Tuesday the 7th day of April 1795.

Voted That Samuel Cornwall be town Clerk.
Voted that Samuel Towner be Supervisor.

Benjamin Haviland,
Nehemiah Jones,   } Assessors.
Stephen Heayt,

David Hickok, Senr.,
&   } Overseers of the Poor.
Jabez Elwell,

Solomon Crane,
Elisha Brown,
&   } Commissioners of Highways.
Abner Crosby,

Abel Hodges, Collector and Constable.
David Barnum, Constable.

Path Masters.

George Burtch, Esqr.,	Joseph Rogers,
Benjamin Lane,	Stephen Yale,
James Birdsell,	Abel Hadges,
Jabez Elwell,	Isaac Crosby,
Daniel Haynes,	Blackleduck Jessup,
John McLean,	Elisha Brown,

Samuel Colwell,
Abraham Mabee, Sr.,
Solomon Fowler,
Abner Crosby,
Jacob Read, and
Elisha Gifford.

Fence Viewers.

Jabez Elwell, Junr.,
John Tweady,
Zachariah Hinman,
Thomas Birdsell,
Abijah Starr,
Elijah Stone,
Roswell Willcox,
David Hickok,
Peter Terry,
Enos Ambler,
Simon Perry, and
Nathaniel Foster.

Pound Masters.

John Tweady,
Silas Burtch,
Roswell Willcox, &
Amos Rogers.

Voted That the next Town Meeting shall be held at the Presbyterian Meeting House.

Voted that the sum of sixty pounds be raised for the maintenance of the Poor of this town."

Patterson Village.—This village, sometimes called Patterson *City*, during the Revolution and previous thereto was called Fredericksburg, and lies in a rich agricultural district in the valley of the Croton. It is about eight miles north-east of Carmel, and one mile south of the Dutchess county line. The Post Office, which formerly was located here, was removed to Haviland's Corner, a little more than a mile east of it, by the Hon. F. Stone, when he was appointed Post Master a few years since. The country in the vicinity and suburbs of the village is charming. The land bears evidence of a neat and enlightened husbandry, with taste in the appearance presented by the houses and their appendages. The gently rolling surface of the land, its freedom from stone, stumps, and bushes—the rich verdure of the fields, and substantial fences enclosing them—the smooth, excellent roads—all combine to make a ride through this portion of the county extremely pleasant and agreeable to a lover of rural scenery. The New York and Harlem Railroad runs between this village and Haviland's Corner.

It was named after the Patterson family, which early settled in the town, the descendants of which are still found here.

Haviland's Corner.—This place is about one and a fourth mile east of Patterson village; and is named after Benjamin Haviland, Esq., who resides there. A Post Office is kept here. On the Post Office Register it is called "Haviland's Hollow." The Hollow is about one and a half miles in length, and one hundred rods in breadth, running east and west.

Towners.—This place was formerly called the "*Four Corners,*" but is now known by the above name, from James Towner, who lives there, and keeps a public-house. A post office and a store is also kept here. Two roads, intersecting each other at right angles, caused it to be called the Four Corners. It is about two miles south of Patterson Village, on the road to Carmel Village.

Cranberry Hill is a small eminence about half a mile east of Judge Stone's residence, over which runs the Birch road. It lies in the east part of the town, and is partly cultivated. Cranberries grew on it; and hence the name.

Pine Island.—This rocky ledge or eminence lies in the middle of the *Great Swamp*, about fifteen rods west of Croton river. This swamp traverses nearly the whole length of the town, and is narrower at the south than at the north end of it. The Island covers about thirty acres of the Swamp, which is about one mile wide. This ledge of rocks rises about two hundred feet above the level of the swamp. It abounds in pines, and hence its appellation.

Hinckley Pond.—This large body of water lying in the south-west part of the town, is one mile long and half a mile broad. It contains excellent perch, pickerel, and other kinds of fish. Its west bank forms the west line of the Harlem railroad. It is named after the Hinckley family.

Little Pond—This sheet of water is in the southeast part of the town, about four miles from the Hinckley Pond, and contains the same kind of fish. It is about half a mile long, and a little more than half a mile in breadth. Its name is the consequence of its being the smallest of the two ponds in this town. The Croton river runs through this town from north to south, and the town of Southeast, ere it receives the main west branch with its smaller tributaries.

REVOLUTIONARY HOUSES STILL STANDING.

1. The old house in which James Patterson, Esq., now lives. It was built by his grandfather, Matthew Patterson, a Judge of the Common Pleas, of Dutchess County, who kept a tavern in it, in the Revolution.

2. The old house now occupied by James C. Hoyt, in Patterson Village.

3. The house now occupied by the widow Dean, about half-a-mile west of Patterson Village.

4. The old house now occupied by Cyrus H. Fletcher, between Patterson Village and Haviland's Corner.

Beverly Robinson, Jun., who was Lieut.-Col. of the Regiment commanded by his father in the British

Army, called the "Loyal American Regiment," at the beginning of the Revolution occupied a farm in this town, which was located in Haviland Hollow, and now owned in part, we believe, by George Stokum, Esq. It was appropriated by the Commissioners of sequestration as a rendezvous for military stores, and keeping cattle, which were collected for the use of the American Army.

"Monday Afternoon, April 21, 1777.
"The Convention met pursuant to adjournment.

Present—Col. Van Cortlandt, Vice-President; Mr. Van Cortlandt, Mr. Harper, Mr. Bancker, Gen. Scott, Mr. Dunscombe—New York.

Mr. W. Harper, Mr. Newkerk—Tryon.

Colo. De Witt, Major Tappen, Mr. Cantine—Ulster.

Mr. Abm. Yates, Mr. Bleecker, Mr. Cuyler, Mr. Ten Broeck. Colo. Livingston, Mr. Gansevoort—Albany.

Mr. G. Livingston—Dutchess.

Col. Williams, Major Webster—Charlotte.

Mr. Smith, Mr. Tredwell, Mr. Hobart—Suffolk.

Judge Graham, Colo. Drake, Mr. Lockwood—Westchester.

Mr. Stevens—Cumberland.

Colo. Allison, Mr. Clark—Orange.

"General Scott, to whom was referred the letter from Hugh Hughes, deputy quarter-master-general, relative to the farm of Beverly Robinson, Junior, reported as follows, to wit: That they are of opinion that, as a very considerable lodgment of stores in the quarter-master's department is formed at Morrison's Mills, in Fredericksburgh, in the county of Dutchess, to and from which there will be much carriage, a proper farm in in its vicinity, for supporting the cattle that may from time to time be employed in that department of service, will be absolutely necessary; and that the farm lately in the occupation of Beverly Robinson, Junior, will be very convenient for that purpose. It is therefore the opinion of your committee, that the commissioners of sequestration in the county of Dutchess be directed to lease the said farm for one year to the said deputy quarter-master-general, at such rent as they shall think proper, notwithstanding any treaty for the same that may have been in agitation between the said commissioners and any individual person, for the use or occupation of the said farm.

"Resolved, That this Convention doth agree with their Committee in their said report."

APPENDIX.

JUDGES OF PUTNAM COMMON PLEAS, FROM 1812 TO THIS TIME.

1812, Stephen Barnum, 1st,
" Robert Johnston,
" Harry Garrison,
" Barnabas Carver.
1813, Joseph Crane,
" Robert Johnston,
" Harry Garrison,
" John Crane,
" Stephen Hoyt.
1815, Barnabas Carver,
" Robert Johnston,
" Harry Garrison,
" Jonathan Morehouse,
" John Patterson.
1818, Harry Garrison, 1st,
" Barnabas Carver,
" John Patterson,
" Jonathan Morehouse.
1820, Abraham Smith,
" William Watts,
" David Jackson,
" John Patterson,
" John Hoyt.
1821, Barnabas Carver,
" Jonathan Morehouse,
" William Watts,
" Abraham Smith.
1823, Harry Garrison,

1823, Barnabas Carver,
" Stephen C. Barnum,
" James Tawner,
" Edward Smith.
1829, Frederic Stone, 1st,
" Bennet Boyd,
" Samuel Washburn,
" Ebenezer Foster,
" Cyrus Horton.
1832, Harry Garrison.
1833, Bennet Boyd, 1st,
" David Kent.
1835, Stephen Pinckney.
1836, Ebenezer Foster.
1838, David Kent,
" Bennet Boyd, 1st,
" John Garrison,
1841, Henry J. Belden,
" Cornelius Warren.
1843, Robert P. Parrot, 1st,
" Azor B. Crane,
" Benjamin B. Benedict,
" Thacher B. Theal.
1845, Nathaniel Cole.
1847, Azor B. Crane, elected Judge and Surrogate under the new Constitution.

APPENDIX.

ASSOCIATE JUSTICES.

1812, Enoch Crosby,
" William Watts.
1813, David L. De Forest,
" Jonathan Ferris,

1813, John Hoyt,
" Enoch Crosby,
" Rowland Bailey.
1815, Enoch Crosby.

LIST OF SURROGATES.

1813, Joel Frost,
" Joel Frost.
1819, Walker Todd.
1821, Joel Frost.
1823, Jeremiah Hine.
1827, Jeremiah Hine.
1832, Walker Todd.

1836, Walker Todd.
1839, Howard H. White.
1840, Abraham Smith.
1844, Azor B. Crane.
1847, Azor B. Crane, elected under the new Constitution.

LIST OF COUNTY CLERKS.

1820, Rowland Bailey.
1823, Jonathan Morehouse.
1826, Jonathan Morehouse.
1829, Jonathan Morehouse.
1832, Jonathan Morehouse.

1835, Jonathan Morehouse.
1838, William H. Sloat.
1840, Reuben D. Barnum.
1843, Reuben D. Barnum.
1846, Reuben D. Barnum.

LIST OF SHERIFFS.

1812, William H. Johnston.
1813, Peter Crosby.
1814, Peter Crosby.
1815, Peter Warren.
1816, Peter Warren.
1817, Peter Warren.
1818, Peter Warren.
1819, Edward Buckbee.
1820, Edward Buckbee.
1821, Joseph Cole.

1822, Edward Buckbee.
1823, Edward Buckbee,
1826, Thomas W. Taylor.
1829, Joseph Cole, 2nd.
1832, Nathaniel Cole.
1835, Thomas W. Taylor.
1838, George W. Travis.
1840, William W. Taylor.
1843, James Smith.
1846, William W. Taylor.

NOTARY PUBLICS.

1816, Peter Warring.
1828, John P. Andrews.
1834, John P. Andrews.
1837, John P. Andrews.
1839, John P. Andrews.

1840, John P. Andrews.
1843, John P. Andrews.
1844, William J. Blake,
" Reuben D. Barnum.

APPENDIX.

LIST OF DISTRICT ATTORNEYS.

1818, Walker Todd.
1821, Frederic Stone.
1829, Jeremiah Hine.
1832, Jeremiah Hine.
1836, Jeremiah Hine.
1838, Frederic Stone.

1841, Frederic Stone.
1844, Frederic Stone.
1847, Charles Ga Nun, elected under the new Constitution.

ATTORNEYS' NAMES.

1812, George W. Niven.
1813, Frederic Stone,
" William Nelson,
" Amos Belden.
1815, Walker Todd,
" Henry B. Lee,
" William Brown,
" John Philips.
1816, Cornelius Master,
" Philo Ruggles.
1817, William H. Johnston,
" Edward Buckbee,
" Moses Hatch,
" Jonas Strong,
" Isaac Hoffman,
" Robert P. Lee.
1818, E. Nye.
1819, James Youngs,
" Stephen Cleveland,
" James W. Oppie,
" Samuel B. Halsey,
" Jeremiah Hine,
" Samuel Youngs,
" J. W. Strang.
1820, Henry B. Cowles.

1836, Stephen D. Horton,
" Lewis Robison.
1839, Eleazer M. Swift,
" Elijah Yerks,
" Howard H. White.
1840, Thomas Nelson,
" Ebenezer C. Southerland,
" Silas H. Hickok,
" Owen F. Coffin,
" William Fullerton.
1841, J. H. Ferris.
1842, Samuel F. Reynolds,
" Benjamin Bailey.
1843, John Curry,
" Charles Ga Nun,
" Charles S. Jorden.
" William J. Blake.
1844, John S. Bates,
" Thomas R. Lee,
" James H. Dorland.
1846, William A. Dean,
" Calvin Frost.
1847, John G. Miller,
" James D. Stevenson,
" Charles M. Tompkins.

LIST OF CORONERS.

1812, William Brown,
" Edward Buckbee.
1813, Edward Buckbee.
1816, Joseph Benedict,
" Samuel Townsend.
1817, Joseph Benedict.
1818, Orrin M. Armstrong,
" David Mooney,
" Henry Holdane,
" Samuel Townsend.
1819, Orrin M. Armstrong,
" Henry Holdane,

1819, Samuel Townsend.
1820, Henry Holdane,
" David Dingee,
" Erastus Smith,
" Orrin M. Armstrong,
" James Dykeman.
1821, James Dykeman,
" Henry Holdane,
" David Dingee,
" William Brown.
1822, Henry Holdane,
" Nathaniel Delavan.

1822, James Dykeman.
1823, Henry Holdane,
" Asahel Cole,
" William Raymond.
1826, Benjamin Dykeman, Jr.,
" William H. Sloat,
" Edmund Burtch,
" Nathaniel Delavan.
1829, William Raymond,
" John Garrison.
1832, Lewis Rogers,
" Seymour Allen,
" David Dingee,
" John F. Haight.
1835, Lewis Rogers,
" John F. Haight,
1835, Stillman Boyd.
1837, James J. Smalley.
1838, John F. Haight,
" Cornelius Nelson, Jr.,
" Abraham Everett.
1839, Amos Tompkins,
" Selah Gage.
1841, Cornelius Nelson, Jr.,
" Jeremiah Dewel.
1843, Addison M. Hopkins,
" James Barker.
1845, Hart Weed,
" Elisha C. Baxter,
" Cyrus Chase.
1846, James Barker.

EXTRACT FROM THE SIXTH CONGRESS OF THE UNITED STATES IN 1841.

Free White Males.

Under 5 years of age	1,001
5 and under 10	860
10 and under 15	792
15 and under 20	755
20 and under 30	1,144
30 and under 40	748
40 and under 50	561
50 and under 60	326
60 and under 70	213
70 and under 80	85
80 and under 90	26
90 and under 100	2
Total Free White Males	6,513

Free White Females.

Under 5 years of age	927
5 and under 10	852
10 and under 15	699
15 and under 20	705
20 and under 30	1,051
30 and under 40	688
40 and under 50	566
50 and under 60	298
60 and under 70	188
70 and under 80	126
80 and nnder 90	39
90 and under 100	5
Total Free White Females	6,144

APPENDIX.

Free Colored Males.

Under 10 years of age	34
10 and under 24	27
24 and under 36	12
36 and under 55	14
55 and under 100	6
Total Free Colored Males	93

Free Colored Females.

Under 10 years of age	17
10 and under 24	24
24 and under 36	16
36 and under 55	8
55 and under 100	9
Total Free Colored Females	74

Female Slave.

55 years of age and under 100	1

Number of persons employed in Agriculture	3,125
Commerce	100
Manufactures and trades	916
Navigation of the ocean	92
Navigation of canals, lakes, and rivers	37
Learned professions and engineers	48
Number of Pensioners for revolutionary or military services	19
White persons—blind	1
Insane and idiots—at public charge	3
At private charge	13
Schools, &c.—Primary and common schools	63
Number of scholars	2,935
Number of scholars at public charge	15
Number of white persons over 20 years of age, who cannot read and write	638

Hotels, Stores, Trades and Professions.

Towns.	No. of Manufacturers.	No. of Mechanics.	No. of Attorneys.	No. of Clergymen.	Total amount of salaries payable to clergymen of all denominations for the year, including perquisites and the use of real estate by them.	No. of Physicians and Surgeons.
Philipstown,	7	325	1	8	3325	4
Southeast,	—	154	—	3	970	3
Patterson,	—	49	1	3	1000	2
Kent,	1	47	—	5	250	1
Putnam Valley,	—	60	—	2	—	1
Carmel,	6	144	3	5	1965	4
Total,	14	773	5	26	7510	15

CENSUS OF THE COUNTY OF PUTNAM, 1845.

Towns.	Total Population.	No. of male persons in the county.	No. of female persons in the county.	No. of male persons in the county subject to military duty.	No. of persons in the county entitled to vote at elections.	No. of aliens not naturalized in the county.	No. of paupers in the county.	No. of persons of color in the county not taxed.
Philipstown,	4209	2175	2034	405	834	170	5	2
Southeast,	2044	1073	971	289	486	39	—	51
Patterson,	1289	617	672	111	308	2	—	24
Kent,	1729	885	844	121	431	5	65	5
Putnam Valley,	1598	833	765	138	370	—	—	3
Carmel,	2389	1175	1214	234	580	6	5	34
Total,	13,258	6758	6500	1298	3009	222	75	119

APPENDIX. 353

Towns.	No. of persons of color who are taxed.	No. of persons of color who are legal voters.	No. of married females under the age of 45 years.	No. of unmarried females between the ages of 16 and 45.	No. of unmarried females under 16 years of age.	No. of marriages in the county during the preceding year.
Philipstown,	—	—	599	268	935	34
Southeast,	22	3	227	191	346	29
Patterson,	1	—	145	153	224	12
Kent,	—	—	209	135	357	18
Putnam Valley,	—	—	212	116	327	35
Carmel,	—	—	298	237	460	25
Total,	23	3	1690	1100	2649	153

Towns.	No. of births in the county during the preceding year.		No. of deaths in the county during the year preceding.		No. of persons in the county born in the state of N. Y.	No. of persons in the county born in any of the N. England states.	No. of persons in the county born in any of the other states in the Union.
	Males.	Females.	Males.	Females.			
Philipstown,	81	87	12	8	3576	60	51
Southeast,	23	22	11	9	1339	166	10
Patterson,	21	14	7	8	1162	114	5
Kent,	33	27	13	11	1702	13	4
Putnam Valley,	35	35	6	2	1567	11	2
Carmel,	42	38	13	16	2217	55	24
Total,	235	223	62	54	11,563	419	96

30*

Towns.	No. of persons in the county born in Mexico or South America.	No. of persons born in Great Britain or its Possessions.	No. of persons born in France.	No. of persons born in Germany.	No. of persons in other parts of Europe.	No. of children between the ages of 5 and 16 years.
Philipstown,	—	479	16	7	9	1103
Southeast,	—	26	—	—	—	484
Patterson,	2	5	—	1	1	319
Kent,	—	9	—	1	—	482
Putnam Valley,	—	6	—	—	1	438
Carmel,	—	18	—	1	—	630
Total,	2	543	16	10	10	3256

Towns.	No. of children attending common schools.	No. of children attending private or unincorporated schools.	No. of children attend'g academies or seminaries other than colleges.	No. of Children attending colleges or universities.	No. of yards of fulled cloth manufactured in the preceding year.	No. yds. of linen, cotton, or thin cloth manuf'd in preceding year.
Philipstown,	646	93	2	—	232	50
Southeast,	448	1	5	2	561	136
Patterson,	240	40	—	—	799	449
Kent,	319	1	—	—	982¼	437
Putnam Valley,	351	—	—	—	1437	246
Carmel,	586	24	1	1	1432½	734
Total,	2590	159	8	3	5443¾	2052

Towns.	No. of yards of flannel & other woollen cloths not fulled manufactured in preceding year.	Acres of land.	Peas.		Beans.	
		No. of acres of improved land in the county.	No. of Bushels raised.	No. of acres under cultivation.	No. of acres of beans.	Quantity raised.
Philipstown,	267	11321	1½	35	6	71
Southeast,	516	21100¾	—	—	¼	3½
Patterson,	982½	12801¾	—	—	—	10
Kent,	780	13441	—	1	1¼	51½
Putnam Valley,	1896	21807	2	26	11	179
Carmel,	769	24067	—	—	2¼	3
Total,	5200½	104,538½	3½	62	19¾	318

Towns.	Buckwheat.		Turnips.		Potatoes.		Flax.	
	No. of acres of buckwheat.	Quantity raised.	No. of acres of turnips.	Quantity raised.	No. of acres.	Quantity raised.	No. of acres.	Quantity raised—pounds.
Philipstown,	566½	7672	40	3350	272¾	5645	¼	39
Southeast,	288½	3503	50	2666	186	17625	½	100
Patterson,	202	2575	42½	1357	145¾	13346	4½	736
Kent,	378¾	4835½	16¾	1196	146¼	8695½	2½	385
Putnam Valley,	804	11483	203	9242	359	10785	5	486
Carmel,	504½	7448	281¼	6695	217	11334	6¼	1086
Total,	2683¼	37,516½	633½	24,506	1326¾	74,430½	18¾	2832

Towns.	No. of acres sown.	No. of acres harvested.	Quantity harvested.	No. of acres sown.	Quantity harvested.	No. of acres sown.	Quantity harvested.
Philipstown,	156¼	156¼	1136	868¼	20466	763	6616½
Southeast,	—	80½	823	776	26871	641	7011
Patterson,	133	267	1420	545½	16692	423¼	3047
Kent,	24	45¼	459	545	13097½	354	2849
Putnam Valley,	34	41	328	823	17488	679	4833
Carmel,	57¼	66	747½	882¾	26244	838	6912
Total,	414½	656	4913½	4440½	129,858½	3698¼	31,275½

Towns.	Iron works.			Trip Hammers.		
	No.	Value of raw materials manufactured in each.	Value of manufactured articles in each.	No.	Value of raw materials manufactured in each.	Value of manufactured articles in each.
Philipstown,	1	$203,310	$397,367	3	$72,692	$128,212
Southeast,	—	—	—			
Patterson,	—	—	—			
Kent,	—	—	—			
Putnam Valley,	1	250	500			
Carmel,	—	—	—			

APPENDIX.

Towns.	Distilleries.			Grist Mills.		
	No.	Value of raw materials manufactured in each.	Value of manufactured articles in each.	No.	Value of raw materials manufactured in each.	Value of manufactured articles.
Philipstown,	—	—	—	2	$7050	$8000
Southeast,	—	—	—	6	13150	14712
Patterson,	—	—	—	3	12884	14172
Kent,	1	$200	$600	5	11300	13230
Putnam Valley,	—	—	—	5	6250	7859
Carmel,	—	—	—	4	8025	9915
Total,	1	$200	$600	25	$58,659	$67,888

Towns.	Saw Mills.		Fulling Mills.			Carding Machines.		
	No.	Value of raw materials manufactured in each.	Value of manufactured articles in each.	No	Value of raw materials manufactured in each.	Value of manufactured articles.	No.	Value of raw materials manufactured.	Value of manufactured articles in each.
Philipstown,	4	$600	$1150	—	—	—	—	—	—
Southeast,	7	1950	2895	1	$1000	$1360	2	$900	$1050
Patterson,	5	655	1350	1	1000	1300	1	1200	1320
Kent,	6	775	1650	2	1687	2227	2	3401	3960
Putnam Valley,	10	1300	2600	1	500	1000	1	500	350
Carmel,	10	1140	2110	1	420	595	1	1240	1440
Total,	42	$6420	$11,755	6	$4607	$6482	7	$7241	$8120

APPENDIX.

Towns.	Oats.		Neat Cattle. &c.					
	No. of acres sown.	Quantity harvested.	No. of neat cattle.	Under 1 year old.	Over 1 year old.	No. cows milked.	Lbs. butter made preceding year.	Lbs. cheese made preceding year.
Philipstown,	768¾	18056	2089	278	1817	1172	96066	100
Southeast,	444	13206	3503	105	1749	1663	178395	10066
Patterson,	464½	10693	2409	198	2211	1016	98625	9344
Kent,	401¾	7328	2567	336	2129	1264	133156	1704
Putnam Valley,	633	13721	2334	277	2024	1094	92157	60
Carmel,	933¼	18412	3181	98	3083	1774	181381	3077
Total,	36,45¼	81,416	16,083	1294	13,213	7983	779,780	24,361

Towns.	Horses.	Sheep.			Hogs.	
	No. of horses.	No. of sheep.	Under 1 year old.	Over 1 year old.	No. of fleeces.	No. of pounds of wool.	No. of hogs.
Philipstown,	425	871	257	611	460	1600¾	1861
Southeast,	372	2027	886	1062	1156	4250½	2594
Patterson,	287	2385	789	1596	1770	5420	1897
Kent,	260	2181	578	1603	1178	3710½	1967
Putnam Valley,	290	1920	834	1126	1114	3599	1736
Carmel,	415	4678	447	4231	2805	10400	2878
Total,	2049	14,062	3791	10,229	8483	28,980¾	12,833

APPENDIX. 359

Paper Mills.

Towns.	No.	Value of raw materials used in each.	Value of manufactured articles in each.
Philipstown,	2	$700	
Southeast,			
Patterson,			
Kent,			
Putnam Valley,			
Carmel,			
Total,	2	$700	

Tanneries.

Towns.	No.	Value of raw materials in each.	Value of manufactured articles in each.
Philipstown,	1	$3500	$6500
Southeast,	1	2500	3500
Patterson,	1	2000	3750
Kent,	2	2450	3940
Putnam Valley,			
Carmel,			
Total,	5	$10,450	$17,690

Deaf and Dumb.

Towns.	Males over 12 & under 26.	Females over 12 and under 26.	Circumstances.—Parents able to support.	Deaf and Dumb of all ages.
Philipstown,	1	1	2	2
Southeast,				
Patterson,	1		1	1
Kent,				
Putnam Valley,				
Carmel,				
Total,	2	1	3	3

Blind.

Towns.	Males under 8 years of age.	Males between 8 and 25.
Philipstown,		
Southeast,		
Patterson,	1	1
Kent,		
Putnam Valley,		
Carmel,		
Total.	1	1

APPENDIX.

Blind.

Towns.	Females under 8 years.	Females between 8 and 26.	Circumstances.—No. whose parents are able to support them.	Circumstances.—No. whose parents are not able to support them.	Total number of blind of all ages.	Males under 21 yrs of age.	Males above 21 years.	Females under 21.	Females above 21.
Philipstown,	—	—	—	—	—	1	—	—	—
Southeast,	—	—	—	—	—	—	—	1	—
Patterson,	—	—	—	—	—	—	—	—	—
Kent,	1	1	1	—	—	3	1	1	1
Putnam Valley,	—	—	—	2	3	1	—	1	2
Carmel,	—	—	—	—	—	—	—	—	—
Total,	1	1	1	2	3	5	1	3	3

Idiots.

Towns.	Circumstances.—No. supported by charity.	Whole number of idiots.	Males under 21.	Males over 21.	Females under 21.	Females over 21.	Circumstances. No. supported by charity.	No. not supported by public or private charity.	Whole No. of lunatics.
Philipstown,	—	1	—	1	—	1	—	2	2
Southeast,	—	1	—	—	—	—	—	—	—
Patterson,	—	—	—	—	—	—	—	—	—
Kent,	—	4	—	—	—	—	—	—	—
Putnam Valley,	—	2	—	—	—	—	—	—	—
Carmel,	—	4	—	2	—	2	1	2	4
Total,		12	—	3	—	3	1	4	6

Churches.

Baptist.

Towns.	No. of churches.	Cost of churches.	Cost of other improvements	Cost of real estate.
Philipstown,	3	800	—	400
Southeast,	—	—	—	—
Patterson,	1	1800	—	—
Kent,	3	4644	50	60
Putnam Valley,	1	900	300	—
Carmel,	2	4900	—	400
Total,	10	13,044	350	860

Episcopalian.

No. of churches.	Cost of churches.	Cost of other improvements	Cost of real estate.
2	4400	150	1000
1	2200	—	—
—	—	—	—
—	—	—	—
—	—	—	—
—	—	—	—
3	6600	150	1000

Presbyterian.

Towns.	No. of churches	Cost of churches.	Cost of other improvements	Cost of real estate.
Philipstown,	1	3000	—	—
Southeast,	1	1500	—	500
Patterson,	1	3000	—	—
Kent,	—	—	—	—
Putnam Valley,	—	—	—	—
Carmel,	2	4940	2500	1400
Total,	5	11,440	2500	1900

Methodist.

No. of churches.	Cost of other improvements	Cost of real estate.	Cost of churches.
6	1400	1050	4800
1	—	100	800
—	—	—	—
3	200	25	2250
2	12	300	2400
12	1612	1475	10,250

Roman Catholic.

No. of churches.	Cost of real churches.
1	1500
—	—
—	—
—	—
—	—
—	—
1	1500

APPENDIX.

Churches. Common Schools.

Towns.	No. of churches.	Quakers. Cost of churches.	Number of common schools.	Cost of buildings.	Cost of other improvements	Cost of real estate.	No. pupils on teacher's list.	Average attendance.
Philipstown,	—	—	12	4295	140	310	567	331
Southeast,	—	—	10	1500	—	—	547	246
Patterson,	1	700	10	1275	—	15	226	151
Kent,	—	—	10	1975	20	—	352	192
Putnam Valley,	—	—	8	1200	100	—	358	—
Carmel,	—	—	12	2647	70	127	374	235
Total,	1	700	62	12,892	330	452	2424	1155

Select Schools. Hotels, Stores, Trades, and Professions.

Towns.	No. of select schools.	Cost of buildings.	No. of children attending.	Inns and Taverns.	Retail Stores.	No. of Groceries.	Farmers and agriculturists.	Merchants.
Philipstown,	4	—	89	10	18	18	130	19
Southeast,	—	—	—	5	6	2	—	10
Patterson,	3	1325	45	4	6	—	208	9
Kent,	—	—	—	2	6	—	143	7
Putnam Valley,	—	—	—	—	3	—	265	3
Carmel,	—	—	—	4	7	8	373	13
Total,	7	1325	134	25	46	28	1119	61

INDEX.

	PAGE
Anthony's, St., Face,	164
Anthony's Nose, Mountain,	167
Altered Taconic Rocks,	30
Ardenia,	176
Augite Rock,	59
Arragonite,	21
Actznolite,	21
Albite,	22
Alumnia,	23
Acicular sulphate of lime,	23
American Precinct, Signers to Revolutionary Pledge in,	128
Acrostic on Gen. B. Arnold,	180
Arnold, Gen., Letter of, to Jefferson,	209
" " " concerning,	197
" Notice of,	190
Asbestus,	20, 21
Areles, Joseph,	85
André, Major, Defence of,	200
" " Letter from,	200
Arsenical iron,	48
Austin Hill,	266
Appendix,	346
Brucite,	21
Basanite,	21
Bradley's Ore-bed,	45
Blunt's Quarry,	52
Beat No. 1, Officers of,	95
" " 2, " "	95
" " 3, " "	95
" " 4, " "	95
" " 5, " "	95
" " 6, " "	95
Beekman's Precinct, Signers in,	113
" " Tories in,	116
Break Neck Village,	160
Bross's Landing,	163
Bull Hill,	163
Break Neck Mountain,	164
Beverly Dock,	204
Brinkerhoof's, Capt., Company,	109
Bedle's, Capt., Company,	110
Boston *Gazette*, Extract from,	197
Barger Pond,	249
Bryant Pond,	249

INDEX.

	PAGE
Berry Mountain,	266
Big Hill,	266
Barrett's Pond,	270
Berry, Lieut. Jabez, Notice of,	277
Barnum, Capt. Joseph, Notice of,	320
Court of Gen. Sessions, Minutes of,	15
Canopus Hill,	248
Chrome iron ore,	21
Copperas, or sulphate of iron,	22
Carburetted Hydrogen,	24
Cotton Rock,	39
Copper and Silver Ores,	46
Crystallized Serpentine,	50
Coalgrove mine,	73
Cold Spring Furnace,	75
Cercome, Thomas,	85
Crane's Mills,	87
Cold Spring Village, Notice of,	158
" " Early settlement of,	149
Continental Village,	162
Cat Hill,	167
Constitution Island,	171
Cat Pond,	178
Collins, bargeman, Notice of	206
Charlotte Precinct, Signers in,	143
Chabasie,	22
Commissioners' letter to Bev. Robinson,	175
Crofts,	247
Clear Pond,	249
Cranberry Pond,	249
Canopus Hollow Creek,	250
Croft, James, Notice of,	227
Carmel, and early settlement of	253
" Extract from Town Records,	256
Carmel Village,	258
Corner Mountain,	266
Cranberry Pond,	269
Crane, Capt. John, Notice of,	271
Crane Family, Record of,	275
Corner Pond Brook,	297
Crane, Joseph, Letter from, to E. Benson,	299
Crane, Thaddeus, Letter from,	300
Coles' Mills,	332
China Pond,	333
Cranberry Hill,	342
Diagram of County,	80
Daton's Mills,	85
Dutchess County, Signers to Pledge in,	102

INDEX. 365

	PAGE
Davenport Family,	149
Duncan, Col., Notice of,	159
Davenport's Corners,	160
Denny Town,	161
Dwight, Timothy,	186
Denny, George, Trial of and Confession,	214
Denny Mine,	72
Drew's Hill,	265
Doansburgh,	292
Dalz Brook,	297
Dick Town,	333
Dean Pond,	334
Epidote,	22
Eel Point,	161
Errrata,	
Flat Rock,	65
Fort Hill,	170
Foundry Dock,	77
Farmers' Mills,	332
Forge Pond,	333
General view of the County,	13
Geology,	17
Ground Ice,	34
Granular Quartz Rock,	31
Granite,	51
Gneiss,	58
Greenstone,	62
Gouverneur Mine,	74
Griffin's, Capt., Company,	112
Griffin's Corners,	160
Graphite,	21
Gilead Pond,	270
Gilead Presbyterian Church,	281
Huestis's Quarry,	37
Hematite Ore-beds,	45
Highland Granite Company's Quarry,	53
Hornblendic Rocks,	62
Heganan's, Capt., Company,	109
Horton's, Captain, Company,	112
Hortontown,	161
Hog-back Hill,	163
Highland Grange,	177
Hyalite,	21
Highland Church and vicinity,	147
Huestis's Family, Notice of,	148
Hemstead's Huts,	247

	PAGE
Horton Pond,	248
Hill, Granny, Notice of,	262
Hazen Hill,	265
Hitchcock Hill,	266
Haviland Corner,	342
Hinckley Pond,	343
Iron Pyrites,	22
Iron Mines,	34
Indian Hill,	265
Johnson, Wm. W., Testimony of,	215
Jefferds, Capt. Samuel, Notice of,	220
Joe's Hill,	293
Kerolite,	21
Kemble Mine,	74
Knickerbocker, Extract from,	188
Kirk Pond,	269
Kent, James, Biography of,	301
Kent, Town of, and early Settlement,	326, 327
" Extract from Record,	329
Laumonite,	22
Lyster's, Capt., Company,	111
Dead Mines,	210
Lake Mahopac,	266
Long Pond,	269
Little Pond,	297
Luddington, Col. Henry,	328
Little Pond,	343
Mahopac Lake,	266
Metamorphic Limestones,	33
Mica Slate,	59
Magnetic Oxide of Iron,	65
Marl, Localities of,	76
Mud Flats,	76
Mead's Dock,	163
Mount Rascal,	163
Muddy Pond,	249
Militia Officers, Return of,	297
Northeast Precinct, Signers in,	122
Nelson's Highlands and vicinity,	151
New Haven *Palladium*, Extract from,	190
Oblong,	99
Orpiment,	22
Oregon,	247
Owen's, Jonathan, Pond,	249
Oakley, Robert, Notice of,	251

INDEX. 367

	PAGE
Philipstown, Early settlement of,	144, 147
Pyroxene,	21
Peat,	21
Pyritious Copper,	22
Primary Rocks,	51
Philips's Quarry,	55
Peat, Localities of,	76
Patent,	77
Poughkeepsie, Signers in,	118
Pine Hill,	169
Philipse, Mary, Notice of,	193
Plymouth Paper, Extract from,	206
Putnam Valley,	247
Pelton's Pond,	250
Peekskill Hollow Creek,	250
Pond Hill,	265
Philips Street Chapel,	280
Pigeon Men, Notice of,	294
Peach Pond,	297
Pine Pond,	333
Patterson, Town of, and early settlement,	335, 336
Patterson, Extract from Record,	340
Patterson Village,	341
Pine Island,	342
Rhinebeck Precinct, Signers in,	134
Roads,	83
Revolutionary Letters, &c.,	92
Revolutionary Pledge,	100
Romans, Bernard, Engineer, Letter from, to Commissioners,	173
Robinson's, Col. Beverly, Answer to Commissioners,	176
Round Pond,	178
Robinson House, Description of,	178
Revolutionary Houses in Philipstown,	209
Richards, Mrs., Notice of,	251
Revolutionary Anecdotes, &c.,	226
Red Mills,	260
Rattle Hill,	264
Round Mountain,	266
Revolutionary Houses in Patterson,	343
Sphene,	22
Schiller Spar,	22
Sulphate of Lime,	23
Sulphur,	24
Sulphate of Iron,	24
Serpentine Rock,	34
Steatile,	44

INDEX.

	PAGE
Stony Point,	54, 169
Sienite Rock,	57
Simewog Vein,	67
Stewart Mine,	71
Southard's, Capt., Company,	110
Sugar Loaf Mountain,	167
Sunk Lot,	171
Silver Mines,	210
Scapolite,	21
Smith, Joshua H., Notice of,	229
Solpeu Pond,	248
Shaw's Lake,	270
Seacord's Pond,	271
Southeast town, and early settlement of,	287
" Extract from Record of,	289
Sodom Corners,	292
Smalley Hill,	333
Talcose Slate,	32
Titanium Ore,	49
Target Rock,	66
Turpikes,	83
Tompkins's Corners,	247
Tinker Hill,	248
Turkey Mountain,	266
Tone's Pond,	296
Towners,	342
Under Cliff,	177
Vinegar Hill,	163
Warren's Landing,	163
Whiskey Hill,	169
Wood Crag,	176
Washington in Love,	192
Warren, John, Notice of,	233
West Point Foundry,	239
Watermelon Hill,	264
Watts' Hill,	265
Wixon Pond,	269
White Pond,	333
Yellow Sulphuret of Arsenic,	22
Zircom,	32

www.ingramcontent.com/pod-product-compliance
Lightning Source LLC
Chambersburg PA
CBHW021134230426
43667CB00005B/110